Towards an Ecology of Language, Communication and the Mind

# *Interfaces*
## Bydgoszcz Studies in Language, Mind and Translation

Edited by Anna Bączkowska

Advisory Board
Jorge Díaz Cintas (Imperial College, England)
Marlene Johansson Falck (Umeå University, Sweden)
Dániel Kádár (University of Huddersfield, England)
Stanisław Puppel (Adam Mickiewicz University, Poland)
Janusz Trempała (Kazimierz Wielki University, Poland)

Vol. 2

Marta Bogusławska-Tafelska

# Towards an Ecology of Language, Communication and the Mind

**Bibliographic Information published by the Deutsche Nationalbibliothek**
The Deutsche Nationalbibliothek lists this publication in the Deutsche Nationalbibliografie; detailed bibliographic data is available in the internet at http://dnb.d-nb.de.

Cover Design:
© Olaf Gloeckler, Atelier Platen, Friedberg

This publication was financially supported
by the Faculty of Humanities,
University of Warmia and Mazury.

ISSN 2195-3368
ISBN 978-3-631-62873-7
© Peter Lang GmbH
Internationaler Verlag der Wissenschaften
Frankfurt am Main 2013
All rights reserved.
Peter Lang Edition is an Imprint of Peter Lang GmbH

All parts of this publication are protected by copyright. Any utilisation outside the strict limits of the copyright law, without the permission of the publisher, is forbidden and liable to prosecution. This applies in particular to reproductions, translations, microfilming, and storage and processing in electronic retrieval systems.

www.peterlang.de

# Contents

Preface ............................................................................................. 7
1  Towards a functional and applicational definition of human language .... 13
2  Interdisciplinary scientific work to understand the essence of the world ............................................................................. 21
3  Ecolinguistics: pathways in research ...................................... 39
4  New perception on *mind*, *meaning* and *cognitive processes* ................ 53
5  Applications and prospects of ecolinguistics as a new linguistics paradigm ............................................................................. 93
References ................................................................................. 113

# Preface

'Science' defined as a communal effort to understand and describe the world (cf. Walach, 2011: 2) starts, nevertheless, from the individual thinker and his/her intuition. Scholarly intuition crosses the safe, commonly accepted thought and research pathways. It usually takes time and effort on the part of this thinker, sometimes it takes next generations of researchers to ground these first visions within the scientifically respected frames. In this sense, science is not always and cannot always be 'mainstream', since it often starts outside the mainstream while still retaining its scientific essence and status. Intellectual and scientific progress can be described by the model:

**mainstream science → scientific outskirts → new science**

*Fig. 1. How shift in sciences progresses*

The turn of the 21st century has brought philosophical and methodological shifts in several sciences. We hear about the emergence of 'new mathematics' (cf. Gleick, 1987: 37). Roger Penrose writes about 'new physics' (cf. Penrose, 2004: 505; Penrose, 2005). Bruce Lipton talks about 'new genetics' (cf. http://www.youtube.com/watch?v=eMlAdEr45dY). The present book is aimed at outlining novel proposals and new pathways in language and communication studies, which leave the mono-disciplinary road of cognitive grammar and cognitive linguistics and enter the multi-disciplinary field of research on man living and communicating within all-that-is – within global/universal contexts. The cognitive revolution of Chomsky's in the 1960's opened linguistics to allow vital cognitive-biological-emotional research into language processes and communication phenomena. Cognitivism in linguistics was the first step towards intellectually satisfying and applicationally potent research in language and communication. We came to new, important understandings; also, we could propose novel, helpful strategies in education, translation studies, public communication, or therapeutic programs. The problem with the cognitive approach is that the research chose the human mind and the collective mind (culture/society) as its ultimate reference points. In other words, for 20$^{th}$ century linguistics, there was not much to study beyond the mind of the communicator (whether the individual or the collective), which was limiting in the sense that the mechanisms of the mind are to a considerable extent based on re-creation and subordinance, rather than creativity and sovereignty. The intuition of the linguistics student or researcher was somehow perplexed at the idea of a man-communicator being so subject to, and one would like to say, hopelessly entangled in the mental stuff of their own and of the collective, mental representation. Now, with the emergence of new world models and new theories of all-that-is, we witness a major step that forefront language researchers are taking towards acknowledging the oneness of the human organism, with the material and non-material constantly pulsating outside. In new linguistics, the human communicator co-exists and co-creates their reality with the surrounding community, using their own cognitive ability, naturally. New linguistics offers new insights into the role of the human mind and its connection to the rest of the context. The mind is no longer the ultimate instance. Human consciousness is not generated by the mind, defined as a self-sustainable, creative machine of thought. The following chapters of this book elaborate on these new methodological perceptions and research assumptions.

Historically, the ecolinguistic pathway in language studies was set up twice, that is by Einar Haugen in 1970, who introduced the metaphorical link between language and language milieu by introducing the metaphor of an ecosystem of language; and in 1990 by Michael Halliday, who redirected the ecological focus to embrace biological and social issues in the language process (cf. Fill and Mühlhäusler, 2001: 43; Steffensen, http://southerndenmark.academia.edu/SuneSteffensen/Papers/1055129/Language_ecology_and_society_an_introduction_to_Dialectical_Lingu).

Fill and Mühlhäusler in their *Ecolinguistic reader* (2001: 43) define these two complementary directions in the following way:

(1) 'ecology' is understood metaphorically and transferred to 'language(s) in an environment' (Haugen, 1972);

(2) 'ecology' is understood in its biological sense; the role of language in the development and aggravation of environmental (and other societal) problems is investigated; linguistic research is advocated as a factor in their possible solution (Halliday, 1992).

Defining ecolinguistics as 'the study of the interactions between any given language and its environment', Haugen (in Fill and Mühlhäusler, 2001: 57) addresses a core assumption of the ecolinguistic stance, which is the dynamic, process-like nature of all language phenomena. Finke writes about the 'many interdisciplinary relations' ecolinguistics embraces in its analyses of language and communication (Finke, 2001: 85). This process-oriented view of language has been taken up by many linguists since the time of Haugen's seminal paper, and the straightforward consequence is that today yet another direction within the ecolinguistic research can be identified. This book is an attempt to propose and discuss the third pathway within ecolinguistics, apart from the Haguen's pathway and the Hallidayan pathway, which can be introduced as follows:

(3) 'ecology' implies holism without reductionism; the language processes are a part of the language ecosystem which itself is embedded within the global ecosphere of the Earth. So, to study language and communication we need a multilayer, grid model of reality, the essential context for any linguistic phenomena.

This volume intends to provide an insight into a new paradigm in contemporary language studies, which will be referred to as ecolinguistics. Its roots are:

1. Haugen's ecology of language (1970s);
2. Hallidayan ecology of language (1990s);
3. cognitive linguistics initiated by Chomsky (1960s);
4. cognitive science as the complex, multiperspective and interdisciplinary science of ourselves and our environment, vibrantly developing in the second half of the $20^{th}$ c.;
5. cognitive ecology, defined by Hutchins as the study of cognitive phenomena in context: not only logical/mental, but also the biological context (cf. Hutchins, 2010).

The theoretical, empirical and methodological support is taken from:

1. quantum models of reality and the mind/brain, in particular the Generalised Quantum Theory (Penrose; Haramein; Walach and Stillfried);
2. the holographic model of the world and mind/brain (Pribram);
3. neurobiological and zoological research into chemosensory communication (years 1990 - 2012).

Supporting research, on which these proposals are founded, comes from other disciplines, and enables a considerably more profound and multi-aspectal analysis of man, environment and 'the life of language'.

This historical introduction is very concise indeed and we will not go deep into the history of philosophy or linguistics to look for the roots of the proposals put forth here. The reader will find the history of linguistic concepts and paradigms in other publications, written with this aim in mind. We believe that the evolutionary force behind scientific progression and paradigmatic change is spiral-shaped. Hence, the concepts, models, visions and assumptions that we seek and point out here never, or rarely, mirror past findings and proposals. It may be but a matter of quality. A subtle shift in some parameters. And the leap occurs which pushes the analyses to another level, previously unexplored. The figure 1 illustrating the process of birth of new sciences, in the four dimensional representation (placing of the object within the physical space and the parameter of time) could be re-build as follows:

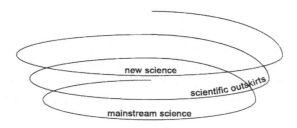

*Fig. 2. Four dimensional model of the scientific shift*

As we perceive it, each reshuffle in the scientific paradigm makes scholars re-define or re-think the conceptual-terminological tools they have so far been using. New linguistics, which is preliminarily delineated in this book and based on the ecolinguistic meta-model, recognizes two major shifts in contemporary science:
1. re-formulation of the model of reality, and
2. re-vision of the scientific method.

As Walach and Stillfried point out (2011: 185), 'our currently accepted scientific model is predicated on a set of presuppositions that have difficulty accommodating holistic structures and relationships and are not geared towards non-local correlations'. Both (i) the reality model which steers the thinking of scientists and people outside science, and (ii) the strategy of doing scientific research consequent on this reality model – that is, the scientific method – are not spacious enough, nor 'aware' enough to embrace the many phenomena, which construct our reality and which are too widely present to be ignored and labeled as 'non-scientific'. Scholars define this sphere not embraced scientifically as 'non-local correlations' (i.e. Walach and Stillfried, 2011), 'parascientific/parapsychological stuff' (Blackmore, 2007; 2009), or 'new physics/new science' (Penrose, 1995). In mainstream linguistics one can hear from time to time unspecified descriptions such as 'language beyond the borders of language' (in Polish mainstream linguistics: język poza granicami języka; cf. Kiklewicz and Dębowski, 2008). This very phrase labels phenomena outside the research interests of traditionally-minded linguistics. However, it also indicates a process that has already started in their minds: first, unspecified, ungrounded intuitions are being born that there is more to language and communication than what mainstream linguistics has delineated as its scope of interest. There is, as ecolinguists see it, no language beyond language. All processes and phenomena which parametrise human communication, on any level of analysis, belong to the language process and are located within the language process. Ecolinguistics as the models' model in the forefront of modern language studies helps to embrace the broad context for language and communication. More than occasionally it means crossing the borders of traditional disciplines and joining forces with specialists of new mathematics, new physics, neurocognitive science, biology, and others. This book has been written to venture a step towards this open, transdisciplinary research plane, where we delineate new linguistics.

# 1 Towards a functional and applicational definition of human language

## Ecolinguistics as the model's model in contemporary language studies

In order to study language behaviours and communication phenomena we have to specify our starting-point definition of language. Linguists, eager to study communication, have to realize and make it clear in their methodological assumptions, how they understand human language as the object of further investigation. The history of linguistics is based on opposing theories and approaches to this central linguistic notion. One model of language is born and, in next to no time, another is proposed as an alternative. In the present analysis, a change in perception will be proposed, namely, it is argued that ecolinguistic proposals, re-born at the onset of the 21$^{st}$ century, are not a reaction against cognitivist models of language and communication, to date dominant in linguistics. Ecolinguistics does not reject the rich tradition of language studies; what it does is an attempt to harmonise them into one concise area of expertise. Ecolinguistics reaches across the boundaries of disciplines to employ helpful methodological, theoretical and philosophical tools, intended to orchestrate all we observe about language and communication into one theory of language. It also re-defines proposals and fills terminological and conceptual gaps in the theory of language and communication. In this sense, ecolinguistics seems to be the metamodel, or the models' model in linguistics. The discussion starts with the fundamental assumptions about what language and communication are. Throughout the analyses we will present new tools which ecolinguistics introduces and which have been proposed in the interdisciplinary dialog engaged in with other sciences. The subsequent sections and chapters outline the consequent extensions to the theory of human language which become possible as a result of implementing new research tools.

## Structuralist and neostructuralist views on human language

If we take into account the scope of 20$^{th}$ century linguistic research and early 21$^{st}$ century linguistic research, language studies have been founded on a basic methodological polarity: research on human language as a system vs. research on it as a process.

The former, systemic approach to language is visible in several definitions collected by Brown (1987), and tabulated below (Table 1.1).

Table 1.1. *Defining human language as a system (after Brown, 1987: 4)*

| What is language? | Source of definition |
|---|---|
| Language is a system of arbitrary, vocal symbols which permit all people in a given culture, or other people, who have learnt the system of that culture, to communicate or to interact. | Finocchiaro, 1964: 8 |
| Language is a system of communication by sound, operating through the organs of speech and hearing, among members of a given community, and using vocal symbols possessing arbitrary conventional meanings. | Pei, 1966: 141 |
| Language is any set or system of linguistic symbols as used in a more or less uniform fashion by a number of people who are thus enabled to communicate intelligibly with one another. | Random House Dictionary of the English language, 1966: 806 |
| Language is a system of arbitrary vocal symbols used for human communication. | Wardhaugh, 1972: 3 |
| [Language is] any means, vocal, or other, of expressing or communicating feeling or thought… a system of conventionalized signs, especially words, or gestures having fixed meanings. | Webster's New International Dictionary of the English Language, 1934: 1390 |
| [Language is] a systematic means of communicating ideas or feelings by the use of conventionalized signs, sounds, gestures, or marks having understood meanings. | Webster's Third New International Dictionary of the English Language, 1961: 1270 |

In Brown's collection of definitions, the main features of systemic definitions are nicely synthesized. In this language as a system approach, a number of vital research assumptions come as a consequence:

1. Human language is perceived as a separate, self-organizing and autonomous system, either vocal, semiotic or gestural in expression; neither endogenic/intraorganismic nor exogenic/interorganismic influences and contexts are involved.
2. Human language enables interpersonal communication as it is composed out of conventionalized elements, agreed upon societally, which, when memorized/learned, are exchanged in the process of communication. Thus, here, human language is, rather unintentionally, reduced to its social and cultural dimension, perceived as a social or cultural process. What limits the approach even more, while pointing to convention and social agreement upon fixed meanings of portions of language, the structuralist approach does not venture to investigate further these cultural and social constraints. The systemic approach to language pays hardly any attention to the fact that convention and cultural/social aspects of language have their dynamicisms as well; we will return to this aspect further in the present volume.
3. Human language defined as a system was not only deprived of any extrasystemic context, but was also treated as a means of communication within the ideal world, and spoken by the ideal speaker. Thus, in this vision of human language Chomskyan performance equates competence.
4. In fact, all elements of the traditional, systemic definitions of human language are vaguely specified and lack applicational potential; we can hardly utilize them when dealing with concrete languages, with real-life communicative situations or in educational settings.

Although it labels itself 'cognitive linguistics' and declares interdisciplinarity and context-relatedness, Cognitive Grammar continues the structural tradition in language studies in a number of aspects (cf. Langacker, 1987; Radden and Dirven, 2007; Górska, 2010). In it, the traditional, structuralist construct of 'langue' is replaced by the cognitive constructs of 'the linguistic representation of the world', and ICMs (Idealized Cognitive Models) are idealized, abstract systems of the conceptualization of reality. The scientific methodology seems to be the same (Kiklewicz, 2006: 50; Bogusławska-Tafelska, 2008a: 53).

Cognitive grammarians study semantic nets which, when activated, construct the link between human cognition and language; they look at language events and derive generalizations on the basis of databases of language samples. In the cognitive grammar model, language is defined as a structured inventory of conventional linguistic units (Langacker in Dąbrowska and Kubiński, 2003: 41). Convention becomes, again, the key notion in this approach. However, cognitive grammarians in their analyses consider two types of language context: they try to include considerations about the convention they notice; and they include the context of the human mind as the map of reality.

Cognitive linguistics, in the Langackerian sense, has to be differentiated from cognitive linguistics understood as a contemporary, interdisciplinary pathway in language studies, incorporating and inspiring such research paradigms as psycholinguistics, sociolinguistics, cultural linguistics, communication studies, and others. So what Cognitive Grammar and Langackerian cognitive linguistics do is modeling a narrow, essential (as it undoubtedly is) fraction of the language process. This model grasps the effects of the human mental system's (imperfect and illusory) adaptational strategy to register communicational and behavioural convention. One of the mental mechanisms in humans is the mechanism designed to grasp and register in a snapshot the short-lived, always illusory, regularity and pattern in the internal/intraorganismic and external reality. This strategy of the human mind results in the, again, illusory yet powerful mental representation of the world, which is based on the illusion of recursion and convention organizing life. In other words, we believe, and agree, that what we see and read from the reality around us and in us, is either conventionally accepted and 'proper', or 'improper' or somehow dysfunctional, since it violates the convention. Values and norms are born at this point, which start to steer human behavior, communicational behavior included. This fraction of our mental activity remains within the scope of interest of cognitive grammarians. In the eye of an ecolinguist, such linguistic approaches, while perceiving language as a system or net and focused on the best model construction to reflect the functions of such a system, fail to account for momentary in principle, often non-rational, based on the individual cognitive system content, unconscious and emotionally marked processes of human cognition. All cognitive processes, together with the language process are to be regarded as beyond cause and effect rationalizations, as on a certain level of processing they exhibit non-local correlations (see Chapters 2 and 3 and 4). Moreover, when one has to deal with a process, its dynamism and cross-contextuality cannot be grasped by models depicting semantic nets (cf. Bogusławska-Tafelska, 2008).

As Patton notes (2002), nonlinear and complex language process becomes a methodological and psychological problem for those linguists who are looking for formal and systemic order. Structuralist and neo-structuralist orientations in language studies have been voicing the intention of some scholars to rationalize, objectivise and put order to language phenomena. Unfortunately, in these approaches so many aspects of the linguistic process and the linguistic milieu are neglected or reduced that structuralism in contemporary linguistics resembles an intellectual game, a logical puzzle, in which players look at certain puzzle arrangements and discuss their function with reference to the whole picture. Here an ecolinguist would point to quantum physics and Heisenberg's principle of uncertainty, according to which when one is trying to measure thoroughly a selected parameter of reality, the other parameters disappear (cf. Lipton, http://www.brucelipton.com/; Kitto et al., 2011: 333). The human mind, in one of its dimensions of activity, resembles a puzzle; however, what we are proposing in this book is that human activity and communicational activity in it are a resultant process of many more mechanisms and dimensions of activeness, which go far beyond 'the mental puzzle'. The next definition of language introduced with the emergence of the cognitivist turnover, opens a new pathway in linguistic investigations and enables the methodological perspective in research to be extended.

## Human language defined as a process

The more we know about human language, its underlying mechanism and the explicit realisation, the more convinced we get of its process-like nature. Next elements of the machinery of human language which linguists, psycholinguists, cognitivists and neurobiologists identify in the course of the research parameterize the very phenomenon. An entity which emerges out of so many co-participating variables has to possess the following characteristics:

multilayer
↕
→ non-linear
→ momentary
→ situationally determined
→ chaotic: 'an orderly disorder' (as Gleick put it, 1998: 15).

The primary feature of human language which is its grid nature brings, as a result, non-linearity, and momentarity of all linguistic and communicational processes. Things happen and pulsate on the level of cognitive structures in the human mind, which provide the cognitive ground for linguistic behaviours. Later, there is pulsation and continual activeness on the level of the structure, in the area of lexicon, grammar, style, phonetics. However rigid and delineated the linguistic formulae are in this regard, these components of the formalized language structure are very much alive and constantly changing. In addition, we can see constant activity on the level of neurobiological basis: from the central nervous system, the endocrinal and genetic systems, to other bodily systems. Further on, there is a continual pulsation in the relation of the organism to the external environment, be it societal, cultural, political, ecological, biological, or other. Later in this book we add yet another layer of processes which co-create life systems and the world around as a macro system – the level of quantum processes.

So, within the cross-disciplinary field of research, human language can be regarded as a process, embracing all the contextualising phenomena from the inside of the organism (the 'endo' perspective) and from the outside of the organism (the 'exo' perspective).

In order to produce language, we have to start language production programs. These are complex programs, which in fact cross the boundaries of the organism of a single communicator (or, otherwise, prove these boundaries be of relative importance). In order to receive language, we start complex language reception programs, which similarly switch on processes crossing the boundaries of the singular living system (for the traditionally psycholinguistic presentation of these mechanisms see Akmajian et al., 1990).

## Conclusion: substance vs. process as the complementary yet incompatible relationship within a whole

Walach and Stillfried in their proposal of the Generalized Quantum Theory as the formal framework for any dynamic, internally and externally complex system, pinpoint that systems are composed of complementary, incompatible observables, which 'are (..) necessary to describe a situation completely, and they cannot be reduced to each other, or expressed in terms of each other nor in terms of a coherent unifying meta-description' (Walach and Stillfried, 2011: 195). These incompatible aspects of a dynamic system are not simple negations of each other; rather, they can be noticed, described and studied on divergent levels of inquiry. The dynamic phenomenon of human language seems to fit this theory (cf. Bogusławska-Tafelska, 2008 a, b). On one level of inquiry, language can indeed be seen as *a system* operating on symbols, being externalized (i) bodily, through gestures, odors or pheromones; (ii) or phonologically or/and graphically, through sounds, grammatical patterns, lexical elements and conversational scenarios. On another, holistic and contextualized level, human language is *a process*, falling into the category of 'life processes', essential to preserve and continue human existence as a species. What we assume in this book, though, is that the traditional view, narrow in scope, on human language as an isolated entity (structuralism), or as a system based on repeatable conventions (neostructuralism or structural functionalism), has to be embedded into and complimented by a holistic, not-reductionist model or rather meta-model. Such a meta-model allows in the rest of the essential context for language and communication.

It is reality as a whole, with its (i) surface processes and objects; and (ii) deep-level holographic/energetic/quantum nature, which is the optimal, essential context for language and communication. Hence, in this search for a method of modelling the linguistic context and outlining a meta-model, we meet researchers from neighboring disciplines. It becomes apparent that we all meet at the task to model or at least notice reality, as in-formated or ex-formated, in Pribram's terminology. Ecolinguistics, as a meta-model proposed here for new linguistics, and more generally, a new paradigm in modern linguistics, meets new physics, new mathematics, new biology, etc. The scientific work over the revised model of the world, and the new scientific method to study reality unifies scholars working on the interdisciplinary field of (new) science. The subsequent Chapter 2 outlines what has recently been done in this field.

# 2 Interdisciplinary scientific work to understand the essence of the world

## The life of language as a quantum phenomenon generated by the quantum mind

The analysis undertaken here in the search for the functional, all-embracing definition of human language brings complications as we go further towards the full context for human linguistic activeness. When we choose a standpoint that language is a dynamic, multilayer process, embedded within the multilevel, pulsating grid of other life processes on Earth (which it is), we come to the point where, in order to locate language and communication within this natural context, we have to include into our methodological framework all the contemporary scientific awareness about the world and its functioning to which we have access today. The complicating aspect is that when we do this, we face the necessity of questioning the framework of the contemporary scientific method used in linguistics. The foundations of the scientific method start to sway because when we let the discussion go freely to allow in the interdisciplinary perspectives, we need to leave the safe, rigid framework of the science of linguistics, and enter the interdisciplinary field of research, in order to discuss the process of language as a quantum process, generated by the quantum, holographic mind.

Del Giudice refers to the recent research advance in physics as 'the revolutions in the physics of the 20$^{th}$ century' (2004:71). Classical physics which, according to the scholar, was present and still developing at the end of the 19$^{th}$ century, was founded on the idea of 'the interplay of two basic entities, particles of matter and fields of forces' (Del Giudice, 2004:71). Matter was built up of separate, independent particles, interacting with each other through forces filling the entire field, i.e. gravitational force, etc. (Del Giudice, 2004:71). The neighbouring scientific disciplines such as biology or the humanities were embedded in the materialistic perspective and developing structure-based, atomistic models and theories. The second half of the 20th century, and especially the recent years, have witnessed a turn towards new understandings of reality. Physics has started to go in the direction of quantum models; biology has started acknowledging non-matter as another quality in the living world; and the humanities have started to acknowledge the wider, environmental (both the internal and external) context for human behaviours and, as a result, have opened themselves to the transdisciplinary cooperation and exchange of ideas and models.

## The holographic structure of the quantum mind

The following section discusses how the contemporary scientific awareness of how reality is constructed affects the foundations of the scientific method in modern linguistics. What we know today about the construction of the world around us and within us, shakes our intuitions and folk knowledge about life and us. After all, we represent the modern, western civilization, very much embedded in our minds which, in turn, have been built upon the western mainstream Newtonian worldview present in educational curricula and supported by mass communication. In other words, newest knowledge and/or speculations about the quantum/holographic construction of reality question the collective conscious / unconscious views on this matter (cf. Bogusławska-Tafelska and Cierach, 2012).

To begin with, the relation between the human brain and the human mind remains one of the re-occuring, unresolved questions in cognitive studies. In the literature, one can find the labels 'mind/brain dilemma', 'the mind/body problem', or 'the Leibniz's gap', which refer to the conceptual gap between mind-oriented studies and brain-oriented studies in science today. It can be noticed that the research outcome in cognitive science is either mentally-oriented and focused on the concept of an abstract notion of 'mind'; or, conversely, is focused on the neurophysiological notion of the human brain, and is biologically-oriented. Cognitivists, generally, avoid the task of translating the knowledge of synapses and neurotransmitters into the knowledge of mental representation, concepts, frames and other mental stuff. What we need at this stage of cognitive research is a cross-disciplinary model which will bridge the research outcome we have collected about the mind and the brain. The distinguished mathematician, emeritus professor of mathematics at Oxford University, Roger Penrose, points out that the automatic, direct equating of the mind with the brain can no longer be accepted. The biology of the brain does not explain the potential of the mind and consciousness (cf. Penrose, 1995: 10). The commonsensical argument against such a unification is that if the brain, as a result of its neurobiological processing, was generating the mind and conscious being, then growing nervous tissue on a laboratory dish would mean 'growing' a new mind and a new 'self'. This has not been done, and seems to be too much of a simplification of the issue. Anyway, mechanistic, algorithmic or biological perspectives are not sufficient or resilient enough to explain the workings of the human mind and the nature of consciousness, understood as a state or process of conscious attention and self-reflection in thought activeness. As Penrose comments on the computational approaches to mind/brain research (1995:10):

> Indeed, electronic circuits are already over a million times faster than the firing of the neurons in the brain (the rate being some $10^9/s$ for neurons) and they have an immense precision in timing and accuracy of action that is in no way shared by neurons.

Thus, there must be other layers of processing which are present in living systems, generate consciousness and enable creative behaviours. The neurobiological layer itself is not the ultimate reference layer here. The conceptual-terminological gap between the mental machinery and the brain machinery needs to be filled by a model which transcends both: the mechanistic, biological framework of brain research, and the cognitivist computational, algorhythmic framework of mentalist studies (cf. ibid.). When we want to simulate on the computer the workings of the human mind, say, during a decision-making process, we will have to take into account, as Penrose puts it, 'a typical outcome', not 'an actual outcome' (1995:23). In other words, computer simulations offer scenarios which are more repetitive and automatic than life scenarios themselves. These are possible or model patterns, however, in real-life settings they hardly ever occur. Scenarios achieved by alghorithmic design and (which equates with computer simulation, in AI research) never develop identically outside the computer lab. In our 2010 research project, we studied aspects of interpersonal communication, the parameters and conditions steering communicative scenarios, and the non-linear, pulsating nature of interpersonal communication. In the results it was visible that one could not possibly foresee the ultimate direction of the communication process, because so many ad-hoc, situational aspects were entering the communication scene that the meaning of the conversation was an inherent mechanism within the dynamic language process, subject to situationally-determined, momentary shifts and changes within the communicative situation, the language process being integral to it (Bogusławska-Tafelska et al., 2010).

Penrose also points to the external environment and its impact as a possible cause for the non-computational nature of human actions, mental activity included (cf. Penrose, 1995:26-27). Environmental modeling – throughout one's lifetime, together with an ongoing environmental influence on the organism and on all the aspects surrounding an organism at each moment – add the non-computational, purely random feature to life processes. Contemporary physics, quantum physics in particular, proposes a less mechanistic view of life and reality than biologists do (cf. ibid.). As we observe today, modern language studies are mechanistic as well. Accepting the full context for any communicative situation, which also means noticing randomness and the non-anticipatory nature of communication, remains outside the mainstream research of contemporary linguistics. The majority of linguists continue to concentrate on the search for systemic regularities; they select language segments to be studied out of a wide situational context. If a discourse or a speech act does serve as the context in a linguistic analysis, usually they represent 'typical' instances, and not 'actual' ones. Very often linguistic analyses resemble intellectual games rather than immersion in 'the life of language'.

Penrose, who left his domestic research field of mathematics to indulge in cross-disciplinary studies, has for several years now been investigating the nature of human consciousness as the central topic in mainstream, world-wide, cognitive science. In his considerations, he voices the standpoint of quantum physicists who, in their visions of the world, go beyond the biological, material model of reality and maintain that primarily the nature of things is non-physical.

> (..) quantum theory seems to tell us that material particles are merely 'waves' of information. (..) Thus, matter itself is nebulous and transient; and it is not at all unreasonable to suppose that the persistence of 'self' might have more to do with the preservation of patterns than of actual material particles (Penrose, 1995:14).

And to continue research into the mysteries of the human mind, as Penrose admits, we will have to broaden what is presently meant by 'science' (cf. Penrose, 1995: 50). What we postulate in this book, as far as modern linguistics is concerned, is (i) a change of paradigm and (ii) several essential alternations within the scientific method.

Before we formulate our proposals for the paradigmatic and methodological frames for ecolinguistics as a new linguistics, we mention here the outstanding contribution to contemporary cognitive science which has been made by the prominent figure of contemporary cognitive science, Karl Pribram. His research is very helpful in grounding the framework of ecological linguistics as new linguistics, as proposed in this volume. Pribram, the world-recognized neurocognitivist, now professor emeritus, has been investigating the human brain for over 50 years. The holographic model of the human brain, embedded in the philosophy of the monistic model of reality, has become central in Pribram's scientific activity. He admits (in an interview: http://www.youtube.com/watch?v=vHpTYs6GJhQ) that the hologram theory agrees in many aspects with quantum mechanics. However, he remarks on the fact that the findings of mathematicians and physicists were not applied automatically in neurocognitive studies; rather, while doing his own research he came across mathematical and physical research and found out that both pathways 'fit' (cf. ibid.; see also Pribram, 2004). Pribram touches here on the very important pattern of how sciences develop and co-operate. Sciences are believed to inspire and set fashions among each other. In fact, though, what we can notice today is blurring the boundaries of scientific disciplines, as traditionally rigid research planes start to expand naturally, being pushed outward by more and more insight obtained into the nature of all-that-is. It is not trends or automatic borrowing among sciences that make them communicate, but the present-day expanding awareness of world mechanisms and phenomena, which leads to the first, preliminary at this stage, assumptions of the same underlying deep structure of everything. What mathematicians, cognitivists and linguistics have noticed and started analyzing independently within the confines of their paradigms, are different aspects of the same whole. This book has been inspired by the presupposition we have had for the last ten years of our academic work that beyond the level of the multiplicity of models, theories, definitions and polarities in language studies, there is a primary level on which contemporary linguists can orchestrate research into one coherent cognitive-ecological theory of ecocommunication (cf. Bogusławska-Tafelska, 2008a.; 2008b; 2011). Having recently come across publications of neurocognitivists, quantum physicists and natural science researchers, we have realized how close we all are to each other in our effort and passion to grasp this underlying machinery of the world. Many of the novel assumptions within innovative linguistics today, result from direct inspiration coming from quantum physics; however, modern linguistics in its most progressive form would not respond to quantum or holographic models if it was not ready for it. This course of events resembles Pribram's account of the beginnings of his research on the holographic model of the brain. He explains that it is not that brain researchers adopted quantum ideas in their work. Rather, at a certain moment, Pribram realized that his ideas about the brain and quantum ideas about the essence of matter (and non-matter) are coherent and touch the same deep level of things. Today, the first linguists are starting to join this multidisciplinary team of researchers.

In order to outline Pribram's monistic vision of reality, it is useful to start here from the Interactive Dualism model built by the eminent neuroscientist, and Nobel prize holder John Carew Eccles. Karl Pribram relates his own model in the context of Eccles' model (cf. Pribram 1. http://www.paricenter.com/center/), which clarifies the proposals. What Eccles put forth is a coherent view on the dualistic problem of the interaction of the mystic mind and materialistic brain. In the brain research, Eccles defines 'dendrons' as being physiological structures of synaptic connections in the brain tissue, allowing brain processes to occur. In mental research, Eccles introduces 'psychons', units of interaction which operate on synapses through quantum processes. In Pribram's study, when one considers the mathematical formulations to describe both the materialistic dendron and the mentalist psychon, one notices that the same formulas describe both. 'The elementary structure of processing in Eccles'material dendron is identical to the elementary structure of processing of a mental (communication) psychon. There is a structural identity to the dual interactive process' (Pribram, 1. http://www.paricenter.com/center/). Pribram (Pribram, 2. http://www.paricenter.com/center/) writes that 'not only radical materialists but also Identity theorists claim that neurological and psychological processes partake of sameness'.

So, we indeed face a dualism of substance and process: mental units of information vs. material units of the physiological brain. This polarized level exists and builds up reality, as Eccles and other dualists maintain. What Pribram claims, however, which has also been very unspecifically touched upon in ecolinguistic research, is that there must be another, primary level of world constitution – Pribram and ecolinguistics use here the same concept of an underlying 'flux' or 'fluctuation' (cf. Pribram, ibid.; 2004; Bogusławska-Tafelska, 2009: 60; 2010: 23, 24). This underlying movement or pulsation gives rise to both matter and non-matter within any given space-time frame. Dualism of matter and non-matter rests upon the monistic in nature potential which itself transcends time and space. This basic level of all-that-is has the form of a hologram, according to Pribram. In the course of his academic work, he has developed his model of reality as a hologram, in which 'everything is everywhere'. It is the lens in the eye, and the lens-like structures in other senses that allow us to receive concretizations of the holographic flux. Similarly to an image on a photograph, humans re-construct reality in their mental representations. Without the lenses of our senses we would receive reality as a hologram (cf. Pribram, ibid.). In quantum terms, very generally speaking, the observer and observed reality are connected and influence each other. The observer models and specifies what is observed. This is the platform on which ecolinguists base their claims for the illusion of objective perception/ judgment vs. subjective perception/judgments (these implications are elucidated and exemplified in the context of communication processes in Chapter 4).

Eccles and Pribram in their analyses provide us with mathematical, physical and neurophysiological insights into the essence of the 'flux'. Ecolinguists continue the work and in research on mind, language, communication and environment they pinpoint the transformation or emergence mechanisms and the consequent output of these transformational process (cf. ibid.). In other words, ecolinguists observe 'the pulsating whole', the flux, out of which further externalizations emerge concretised by their space-time parameters. The space-time framework as investigated by physicists and mathematicians, is analysed further on as studied in local research; in linguistic studies, it is subdivided into many more parameters, which serve to describe and analyze communicative situations, discourse parameters, situationally-determined communication meaning, the system of language itself, etc. Hence, each linguistic model, paradigm and perspective in fact touches upon and investigates some parameters of the 'exformation' (externalized, palpable form of flux), or 'information' (internalized, minded form of flux) (cf. Pribram, 1, 2, http://www.paricenter.com/center/; 2004). Examples of the first category belong to the language production phase: the articulatory or textual features of the language process and communication; while examples of the second category include all 'mind stuff' behind linguistic activity, i.e. cognitive representations of reality, mental processes like problem-solving or evaluations, thoughts, etc.

## Quantum brain dynamics

Pribram explains his quantum model of brain activity in the context of Roberto Llinas' model of neural brain processes. These two models complement each other. Llinas' model analyses circuits and the firing activity of neurons, and his spectrum of interest refers to – as Pribram names it, having borrowed the terms from Chomsky – surface structures. What Pribram models in his work, conversely, is the deep structures of brain work, which involve fine fibres of the neurons and the quantum processes that control the activity of these fibres (cf. Pribram 3 http://www.paricenter.com/center/). 'Brain process is essentially a search process: the brain, conditioned by earlier experience, searches for a satisfactory response to the new situation that the organism faces' (cit. Pribram, ibid.). The search process itself is modeled by both Llinas' and Pribram's models, complementarily. The proposal of brain/mind activity being a quantum activity solves two baffling problems: the binding problem (cf. Bogusławska-Tafelska, 2008), and the time constraint, which refers to the brain deciding on good solutions and choices without time expenditure, even when the situation presents a multitude of potentialities. The quantum state of superposition, when all possibilities are equally available and can be explored simultaneously, solves here the puzzle of how the brain/mind does this (cf. ibid.).

In this very sketchy presentation of mathematical, physical and neurocognitive accomplishments, it is vital to note that science has approximated spirituality as a traditionally non-scientific domain of human experience; even concepts and terms used by Pribram and others in their scientific discourse pave the way for a new type of communication, i.e. communication including the concept of waves of energy, vibration providing energetic background for brain activity, or the idea of a multitude of potential choices available at the problem space of the problem-solver; these have traditionally belonged to spiritual discourse. We will elaborate on this further in the subsequent sections of this chapter.

## Walach and von Stillfried's model of Generalised Quantum Theory to describe behavior of all living systems

Harald Walach and Nikolaus von Stillfried (2011) undertake the task of modeling the behavior of all living systems on the basis of what quantum theory proper says about quantum systems. They maintain that, if generalized enough, the quantum theory can apply to all these systems, which so far have been treated with suspicion as 'unscientific', and neglected, regardless of their wide occurrence (see also Filk and Römer, 2010). Specifically, this model embraces all unstudied cases of non-local relatedness, when there are no casual signals to maintain the relation, yet the relation exists. In its standard strategy, the scientific method is based on the general idea of extracting replicable, material/tangible casual signals of the systems it studies. Walach and Stillfried point to a vast collection of systems which are internally related and exhibit relational features, but there are no casual signals between them (2011). 'There is only relatedness, but no signal' (cit. Walach and Stillfried, 2011: 191). The collection of such phenomena with the reproducibility problem includes all so-called parapsychological phenomena which have a rich anecdotal documentation, among them being telepathic communication, clairvoyance or precognition (cf. ibid.). All efforts to replicate these systems in order to demonstrate their occurrence, using standard experimental procedures, have failed. Walach and Stillfried notice that in order to study these systems as a vital part of 'transhistorical and transcultural' human experience (cf. ibid.), one has to use other than standard scientific tools. Generalised Quantum Theory helps to shed new light on these mechanisms. In Chapter 4, this model of system behaviour is applied to re-define and re-consider a number of traditional linguistic key notions, i.e. the idea of linguistic/conversational *meaning*, the notion of *the human mind*, or *human cognition* and its illusions embodied and sanctioned, or the illusion of manipulation by means of language.

## The interdisciplinary paradigm in contemporary cognitive linguistics

A paradigm is a coherent world view (cf. Patton, 2002: 71). It is defined as 'a patterned set of assumptions concerning reality (ontology), knowledge of that reality (epistemology), and the particular ways of knowing that reality (methodology) (cit. Guba in Sale et al., 2002: 44; cf. Cibangu, 2010). Hence, paradigms are hidden behind research methodologies, or simply behind particular modes of reasoning, comprehending, and/or mediating reality.

In order to allow into linguistic research all the present, valid, interdisciplinary research material describing the world, man and communication processes, we need a new paradigm in language studies. Ecolinguistics studying language and communication processes has become our proposal for present-day linguistic research. At this point in our discussion, alternative, new pathways in 'eco' linguistics as new linguistics are pointed out. Implications and consequences which come together with a fresh, 'eco' view on language and cognition fit very naturally with what quantum physicists or holographic theory proponents say about their understanding of the nature of the world.

What Pribram names in his studies as 'ontological monism' (cf. Pribram, 1. http://www.paricenter.com/center/), we refer to as 'holism without reductionism' (Bogusławska-Tafelska, 2008a, 2008b).

Table 2.1. *Fundamental methodological and philosophical shifts to be considered when entering the ecolinguistic paradigm in contemporary language studies*

| Classical model of the world sustained by mainstream education and collective consciousness - in the western tradition | Alternations within the model of the world, as induced by recent, cross-disciplinary research outcomes |
|---|---|
| 1. Classical Galilean, Newtonian and Darwinian visions of the nature of reality:<br>a. nature of the world rests upon matter, studied by such sciences as traditional biology, chemistry, traditional physics, western medical science, etc.;<br>b. according to many cognitive scholars, consciousness is always associated with nervous activity in a complex brain; by changes in sensory input, by injury, disease or drugs one can alter conscious states; the view has it that matter generates non-matter;<br>c. other voices in the mainstream cognitive science today consider human consciousness in the context of activated schemas and frames in the abstract mental representation (hundreds of them, active at any time to subserve the current interaction of the organism with its environment). Hence, the preconception is that there are two streams of research: | 1. Visions of the nature of the world consequent on findings of quantum mechanics, mathematical, biological and neurocognitive research:<br>a. field of energy – or 'non-matter' – which generates 'matter', with the mandatory presence and participation of the observer; hence, the observed is parametrised by the receiving mechanisms of the observer;<br>b. the nature of things is 'non-physical' primarily (Penrose, 1995; Schrödinger, 2007; Patton, 2002; Pribram, 1, 2 http://www.paricenter.com/center/; 2004);<br>c. the notion of 'self' and 'consciousness' are to be studied within the non-matter, energetic framework;<br>d. preservation of 'self' may have to do with the preservation of patterns rather than of actual material particles (cf. Penrose, 1995:14);<br>e. according to the holographic theory, the primary aspect – the energy field or non-matter – constitutes a deep structure in the construction of reali- |

| | |
|---|---|
| mind-oriented research, and body-oriented research; the dualism escapes any theoretical reconciliations. | ty. On this underlying deep structure polarities, dualisms, or conceptualizations do not exist, as everything is fluid and interconnected; |
| | f. Pribram notices that were we deprived of the lenses of our eyes and the lens-like processes of our other sensory receptors, we would become immersed in holographic experiences (cf. Pribram, 1, 2 http://www.paricenter.com/center/; 2004). |

## Towards reconciliation of the schism between the materialistic/scientific domain and the spiritual domain of human experience

The ecological proposals in modern humanistic and interdisciplinary research go so far as to allow the scientific explanation or insight into treated until now as 'spiritual' aspects of reality. What we claim here was also noticed by scholars of divergent research backgrounds, among them Pribram, Schaefer, Haramein, or Walach and Stillfried (cf. Pribram, http://www.paricenter.com/center/; 2004; Schaefer, 2006; Haramein in http://juicylivingtour.com/2011/10/25/the-unified-field-nassim-haramein/; Walach and Stillfried, 2011). They point out that this quantum/holographic/ecological vision of the world is starting to deal scientifically with mystical/spiritual insights, rationalizing them and grounding them in scientifically accepted formats. New sciences cross the boundaries of traditional scientific methodologies used to study man and reality; the conceptual-terminological fields allow in the concept and term 'spirituality', referring to the underlying, energetic, pre-conceptual basis of life (spirituality, however, is not to be confused with the concept of 'religion').

In fact, the definition of spirituality proposed by Walach (2011) nicely co-exists with contemporary definitions of perception and cognition. Walach writes (2011: 6):

I define spirituality as an experiential realization of connectedness with a reality beyond the immediate goals of the individual (..). It gives rise to a holistic type of knowing that manifests cognitively, emotionally and motivationally. This is why it is termed 'experience' in the sense of an inner experience of reality.

Thus, spirituality refers to perception of reality through experience. What is essential and differentiates this type of perception from cognitive perception is its holistic scope. Spirituality means perception through experience which employs not only cognition, but also emotions and motivations (cf. ibid.). In its wide scope, spirituality as experience transcends the individual perspective and touches the universal level of all-that-is.

This definition grasps the essence of experience and perception as seen by ecologically-minded scholars. The holism highlighted in this definition refers to the fact that humans perceiving and experiencing the environment around and in themselves, activate their multimodal, cognitive, neurobiological, linguistic and non-linguistic resources in the very process. The well-recognized Polish ecolinguist Professor Stanisław Puppel sees in this way the human interaction with the 'endo' habitat and the 'exo' habitat (cf. Puppel, 2011). Furthermore, in the spiritual type of perception and experience humans use their overall classical and quantum tools; it is not sense organs exclusively that are working, but the organism in its completeness, which generates personalized consciousness able to receive and experience life. We can say that it is consciousness which is involved in spiritual experience (cf. Walach, 2011).

As Walach and Stillfried notice (2011: 195), the domain of science and the domain of spirituality form a complementarity relationship and together describe reality as a whole. The reason that 'spirituality' has not been academically accepted as a scientific topic for a scientific study, thus being 'incompatible' with 'science', is that traditional world models and the scientific method itself have not been holistic enough to incorporate a host of phenomena which are a part of our life, yet marginal or outside the studied part of it. A new paradigm and a revised scientific method will not only allow a holistic view on existence and on humans, but will also provide methodological tools to scientifically deal with phenomena which Walach and Stillfried name as instances of 'non-local relatedness' (2011). They say that

> the current paradigm (which is descending now – M.B-T.) does not really allow for non-locality outside of quantum physics. This is related to the assumption that any regular relatedness can only be mediated via material casual signals (..). This situation is the reason why all instances of non-local relatedness, such as telepathy, clairvoyance, precognition or telekinesis, have been viewed with suspicion by mainstream science (2011: 190).

They add that

> (..) if these phenomena exist (parapsychological, non-local – addition M.B-T.), they are likely not of a casual nature in the standard sense. At the same time there is such a wealth of anecdotal and phenomenological evidence that it would be unscientific to deny the fact that these phenomena are a transcultural and transhistorical part of human experience (2011: 191).

Walach and Stillfried in their recent paper propose a new framework for studying the behavior of any system, beyond the existing systems. Generalized Quantum Theory is a framework which contains a model of generalized non-locality (see previous sections of this chapter). In effect, what we are offered is a tool to study a larger portion of the stratified reality, than the one that could be embraced by traditional scientific disciplines.

In academic publications on divergent aspects of reality, not only in mathematics and physics papers, but also in psychological or psychotherapeutic research, we can find numerous references to the term 'spirituality', and to 'spiritual' models of the world. For example, Sills and Lown in their paper on subliminal processes and psychotherapy (2008), very eagerly adopt the terms and ideas of Buddhist psychology. They openly admit that Eastern philosophical and religious traditions contain useful information and tools for modern cognitive sciences; in their analysis they use a innovative and paradigm-crossing term from Buddhism namely, 'the mind-body', which is related to the human as the cognitive – physiological whole. Having borrowed this term/concept, the authors of the paper seem not even familiar with the novel findings of the neighboring sciences i.e. physics and mathematics. Again, as in the case of modern, forefront linguistics, the field of psychotherapeutic and cognitive research is internally driven to accept an extended, multispectral model of reality and man, and is starting to notice and propose an extended context in the research.

While academically-based physicists and mathematicians spend years over hard physics to produce rigorous, precise calculations and scientific documentation, cross-disciplinary dialog can be ventured due to 'bridge builders', that is people who have taken to task of translating and simplifying 'hard findings' of their colleagues. Einstein, Podolsky, Heisenberg or Feynman, to mention but these top names in physics, have to be translated to representatives of neighbouring fields of research. Even if what we take from this research is but the inspiration to formulate new scientific metaphors. As the Swiss-born self-trained physicist Nassim Haramein observes (http://juicylivingtour.com/2011/10/25/the-unified-field-nassim-haramein/), the non-material/spiritual world defines and gives birth to the material world; it is time to write the physics for it. The scholarly work to ground these philosophical and ancient-traditional ideas into modern science has busied Haramein for decades, as he admits. Today, Haramein is becoming recognized in international physicist circles, and is awarded for his papers at international physicist conferences. He is introduced as a physicist who is crossing paradigms.

## Modern physics: definitions of consciousness, and transcending the perspective of the individual towards the unification of all-that-is

In his years-long research, Haramein has been looking into the outcome of not only classical physics but also chemistry, biology, ancient philosophy and history. In his scientific papers, he uses his equations and mathematical/physical analyses to describe the nature and construction of the world, consciousness, and related notions. We collect these proposals in the table below, since they have a direct bearing on what we say in this book about the ecolinguistic stance in modern linguistics. Hard physics and mathematics have enabled to extract the basic logic of quantum mechanics and quantum theory (cf. interviews with Nassim Haramein found on the internet sites http://juicylivingtour.com/2011/10/25/the-unified-field-nassim-haramein/ and http://theresonanceproject.org/):

Table 2.2.  *New physics and new linguistics: philosophical and methodological assumptions*

| Topic or issue in the focus of cognitive science today | Newest findings within physics and mathematics (on the basis of Nassim Haramein's research published or forthcoming) |
|---|---|
| **consciousness** | 1. Consciousness cannot be found in the brain. The physiological brain is, metaphorically, like a radio set, functioning as a resonance device. The announcer is elsewhere. |
| | 2. We are submerged within the unified energy field that connects everything, and that is the source of all matter and non-matter. |
| **the building substance of the universe** | 1. The inside of the atoms composing molecules is 99.999 % space. |
| | 2. The space between planets, the space between the atoms, or the space inside the atom is not empty. It is full. It contains energy, i.e. the source energy of everything. In quantum physics it is referred to as 'vacuum fluctuation'. |
| | 3. The amount of energy within the atom is infinitely dense. The amount of energy inside the proton equates to the exact mass of the universe. |
| | 4. Hence, all components of the universe are holographically expressed within one atom. |
| | 5. If an atom is 99.999 % space filled with energy, we and all matter around us are energy as well; or, in other words, all matter comes from non-matter, is built out of non-matter. |
| **holographic model of all life** | 1. We are composed of atoms, which holographically represent the universe. Hence, we ourselves are holograms of the universe as well. In other words, being components of the world around us, in ourselves we contain the information and reflection of the world. |

| | |
|---|---|
| **natural energy / life resources** | 1. Humans and all-that-is are embedded and naturally plugged to an unlimited energy resource, which is the 99.999 % energy fluctuating inside the vacuum of material particles composing matter. |
| | 2. In order to tap into the infinite source energy, we need to go inward; this strategy is both based on modern mathematical and physicist calculations, and on the philosophies of most great thinkers throughout the history of humankind. In other words, modern mathematics offers calculations and models to ground spiritual and philosophical intuitions in modern scientific frames. |
| **Information transmission, motivation, environmental impact** | 1. The unification of all-that-is is a consequence of the mathematical equations proposed. As Haramein notices, once familiar with this model of reality, one automatically becomes humanitarian in one's actions. There are no autonomous systems. People together with micro and macro objects of the world are all interconnected, as they are formed from and embedded in the same source energy. |
| | 2. A person's cognitive, emotional or physiological activeness always occurs in the context of the unified field of universal energy and its material embodiments. On the one hand, we are – quite literally – able to create on the material plane what we wish to create; on the other hand, however, all our mental, cognitive or physiological actions are connected with the activeness of the environment, understood at all possible scales: from micro to macro. |

# Conclusions

The advantages and benefits of transdisciplinary dialog and support in current scientific research are mutual: modern physics and mathematics provide cognitive sciences with the calculations and axiomatic formulations to ground new visions and new models within a scientific, formalized framework. It is no longer intuitions or unsubstantiated divagations but science documented by means of equations and empirical research. On the other hand, cognitive sciences, and ecolinguistics at the forefront of modern communication studies, with their readiness to absorb and further the mathematical models of reality, add further validation to these models.

In the next chapter, the ecolinguistic stance in modern language studies is presented. We focus on the main assumptions of the ecolinguistic paradigm; on the research which raises the fundamental issues of the essence of reality and its dynamic makeup; and on the empirical research done within this novel linguistic framework.

# 3 Ecolinguistics: pathways in research

> *Knowledge should not continue to suffer a conceptual splitting between human science and natural science.*
> *(G.G. Globus, K. Pribram and G. Vitiello, 2004)*

## Introduction

As Alwin Fill, the well-recognized researcher in ecoliguistics today, writes, 'ecosystems are life systems' (2001: 45). Regarding human language as a living attribute of life on Earth is usually treated metaphorically. In this chapter we would like to point to scientific evidence which goes beyond metaphor into new physics and new mathematics and studies the life of language in its naturalistic, not necessarily metaphorical sense. We shall start with a concise outline of the history of the ecological stance in modern linguistics.

## Ecolinguistics: the first opening

The conceptualization of language as a living organism is attributed to the German linguist August Schleicher (Drogosz, 2010: 64). 'He explicitly compared languages to evolving species and represented genealogical relationships among languages in the form of a tree (..) (ibid.: 64). According to Drogosz, the metaphor of 'language as a living organism' is still central in the present ecolinguistic publications of Mühlhäusler, Steiner, Mackey, Haugen and Puppel (cf. ibid.). As regards the introduction of the ecological point of view and the theory of ecosystems to the study of language, this happened somewhat later, through the Norwegian-American linguist Einar Haugen.

It is essential to notice two openings for the ecological approach in modern linguistics. Haugen, in 1970, initiated the first opening. He put forth a concise, very futuristic definition of ecological linguistics, which for him was the study of interactions between any given language and its environment. (..) part of its [of language – M.B-T.] ecology is therefore psychological: its interaction with other languages in the minds of bi- and multilingual speakers. Another part of its ecology is sociological: its interaction with the society in which it functions as a medium of communication (2001: 57).

Haugen also introduced the key metaphor of ecolinguistics which is 'life of language':

> In writings of the nineteenth century it was common to speak of the 'life of languages', because the biological model came easily to a generation that had newly discovered evolution. Languages were born and died, like living organisms. They had their life spans, they grew and changed like men and animals, they had their little ills (..) (2001: 57-58).

As Finke observes, soon after the theory of ecosystems was introduced into linguistics, linguists and cognitivists noticed the materialist orientation of the biological models, which was an obstacle in the research into the mental aspects of language, mind and communication (2001: 84-85).

Around that time, Gregory Bateson voiced his ideas of 'ecology of mind'. He saw the human mind as a part of nature and natural ecosystems, not outside of them which in itself was a breakthrough proposal (cf. Finke, 2001: 85; Bateson, 1996). For Bateson the mind as a part of nature is a reflection of nature and the natural world. In his book *Mind and nature. A necessary unity* he considers a holistic basis for all life which is realized in the common pattern or rule behind all creation (cf. Bateson, 1996).

Table 3.1. *Consecutive phases of the development of the 'eco' perspective in the contemporary linguistic studies: Steffensen's outline (cf. http://southerndenmark. academia.edu/SuneSteffensen/Papers/1055129/Language_ecology_and_ society_an_introduction_to_Dialectical_Lingu)*

| time span | trends within 'eco' linguisitics |
|---|---|
| 1970s and 1980s | Haugenian approach realized within such fields as: language acquisition, bilingualism and multilingualism, language diversity, language death, cognitive research and sociolinguistic research; |
| late 1980s and early 1990s | focus on the biological environment; on environmental and societal problems and ways to deal with them; |
| through the 1990s | ecolinguistics developed as an autonomous field of research within applied linguistics. |

The early years of ecolinguistic research were directly inspired by biological models and, subsequently, by social models. As Fill notes: 'the idea of transferring concepts, principles and methods from biological ecology to the study of language was soon extended by a group of German researchers (..) in an approach called 'ecological linguistics' (2001: 44-45). The first ecolinguists were interested in developing either the biological idea of ecosystem as applied in language studies, or in the metaphorical extensions of the concept 'the ecosystem of language'. In the 1990s, when Halliday proposed his views on growthism, sexism and classism in language, social models and theories were also infused into the newly-born linguistic sub-discipline (cf. Halliday, 2001). This starting-point infusion from the natural and social sciences gave linguistics a new momentum, which paid dividends several years later when this cross-disciplinary and cross-paradigmatic trend gave rise to fully sovereign and independent scholarly explorations of the linguistic phenomena embedded in an essentially wide context, both external (exogenous) and internal (endogenous). From the time of the first direct absorption of such inspiring biological and social ideas, ecolinguistics developed on its own, and 'the second opening' within this paradigm, which came around the year 2000, was a different, more autonomous process.

Although there are voices maintaining that ecological linguistics has been growing ever since Haugen's insightful proposals (i.e. Steffensen, 2007 – http://southerndenmark.academia.edu/SuneSteffensen/Papers/1055129/Langu age_ecology_and_society_an_introduction_to_Dialectical_Lingu), it seems that the 1980s and early 1990s were the time when the ground was not ready for ecolinguistics to gain a second momentum, in the sense of both the methodological tools being absent, and in the sense of the scholarly awareness not being ready to absorb the wide context for language processes. This is why it seems reasonable to divide the time span of the last 40 years, since Haugenian speech, into two periods, the second of them embracing the last 10 - 15 years; this second opening marks a sovereign and novel research trend within the new paradigm.

## Ecolinguistics: the second opening

The second opening for the ecolinguistic paradigm in modern language studies is still a fresh phenomenon; it is still in process. While the Haugenian opening resulted from direct inspiration from the biological and social sciences, the recent 10 to 15 years have witnessed a different scenario. The second momentum in ecolinguistically-oriented studies has been gained as if from within the linguistic research itself. Linguists have been inspired by the ecolinguistic policy of holism in methodology, and the multilayer treatment of all cognitive/biological/sociological processes which are the context for language, and as a result of this freedom from the structural framework of traditional linguistics, novel and insightful points of view have been put forth. Below, we tabulate several examples of linguistic postulates, which are representative of the second opening of the ecolinguistic paradigm. Interestingly enough, some of them have not necessarily been intended to represent this new perspective. Their authors – linguists, cognitivists or interdisciplinary researchers – have been studying the communicative, neurocognitive, cognitive, social, psychological, biological, mathematical, and ecological conditions of human activity, and have quite naturally arrived at models and research assumptions which have eventually moved ecolinguistic research up to yet another philosophical and methodological platform. The list cannot, though, be regarded as complete. The names and publications included in it are representative of some thought pathways and new insights.

*Table 3.2. Inspirers of the re-introduction of the 'eco' approach in cognitive science in general, and in ecolinguistics specifically*

| cognition: processes and parameters | philosophical, methodological and applicational proposals which delineate ecological linguistics today |
|---|---|
| human consciousness and human mind | 1. Susan Blackmore investigates human consciousness: *Consciousness. An introduction* – first published in 2003, and *Conversations on consciousness* – first published in 2005.<br><br>Human consciousness is presented from a multitude of methodological and philosophical perspectives. Blackmore invites researchers representing divergent disciplines; she initiates an interdisciplinary dialog between cognitivists, philosophers, biologists, mathematicians and neurologists.<br>She draws on eastern ancient philosophical-religious traditions of Buddhism, Hinduism and Sufism to talk about such 'fringe' phenomena as meditation, illusion of reality, altered states of consciousness, etc.<br>2. Karl Pribram models the human mind. The mind is seen as a hologram. This model allows several so far unresolved issues in neurocognitivism to be approached such as: the binding problem, mind-body dualism, or the relations between the individual self and the collective self as founded on the quantum nonlocality model.<br>3. Roger Penrose in his studies on the nature of the human mind and consciousness applies concepts from quantum physics and new mathematics. His work resulted in new understandings of the intraneuronal site of creativity and consciousness; new explanations for the mind-body dilemma, and quantum non-locality and its implications in mind/brain research. |

| | |
|---|---|
| non-linearity and pseudorandomness of the life processes, human cognitive processes included | 1. James Gleick in *Chaos* (1998) promotes and explains a new vision of life as being based on the idea of pseudo-random chaotic processes. As Gleick observes, chaos refers to 'an orderly disorder'. The activity of the human mind is submitted to this chaos model of the world as well. |
| | 2. Michael Quinn Patton in his book on qualitative research in modern science (1990, 2002), successfully undertakes the task of inlaying cross-disciplinary new models of reality into the mainstream scientific methodology. He uses such concepts as holistic treatment of world phenomena beyond dualisms, non-linear dynamics, a systems perspective on the pulsating nature of life, Gleick's chaos model, and others. |
| the underlying primary substance as the unifying basis of life on Earth | 1. Gregory Bateson in *Mind and nature: a necessary unity* (1979) sees the inter-relatedness of all elements of the natural world. He points out common patterns which link aspects of matter and non-matter. He identifies a meta-pattern which orchestrates the natural world |
| | 2. Peter Mühlhäusler in his publications, in accordance with the ecolinguistic postulates, proposes re-defining a number of traditional linguistic notions to locate them within the linguistic ecosphere (Mühlhäusler, 1995; 2003; Fill and Mühlhäusler, 2001). |
| | 3. Stanisław Puppel in his ecolinguistic publications remains consistent in relating linguistic phenomena to one underlying, unified base. He investigates the grid of the semiosphere, embedded within the Earth biosphere. In his recent publication (2011), Puppel introduces the concept of General Mechanism of Linking, operating on the commonly shared, fluctuating level of the communication network, across the human species. Communication occurs within various layers, verbally, and non-verbally by activating different non-language recources. |

Ecological points of view have been brought to linguistics together with the assumptions put forth by those environmental writers who see linguistic and philosophical reasons for the ecological crisis (cf. Mühlhäusler, 2003: 2). 'Green language' and 'green discourse' are indeed a part of ecolinguistic research today. What has to be noted, though, is that contrary to layman intuitions and public awareness, the environmental campaign and consciousness raising about 'being green' are not the only concerns of ecological linguistics. Ecolinguistics, with its regard to ecosystem and context for language and communication, goes deep to the core of not only modern linguistics but modern science in general, and relates to (i) models of the world, and to (ii) the scientific method itself. When we apply ecological thinking to science, we eventually incite a methodological and philosophical reshuffle. This study is intended to go deep to these levels as well, and report on the fundamental shifts we witness once ecological views have been infused into the social and natural sciences.

## The primary, unifying substance of the world in the eyes of language researchers

Throughout the history of modern science and across scientific disciplines, a growing number of researchers have been recognizing and investigating the primary source of life and life processes. At the threshold of the $3^{rd}$ millennium, scientists are more capable then ever before of organizing the partial research findings, done in isolation of traditional autonomous scientific disciplines, into one coherent theory of all-that-is. Today, in the scientific work worldwide, we notice a mosaic of models and rules, harmoniously co-voicing the information about how reality works. In Chapter 2, mathematical and physicist calculations were introduced into the discussion. In this chapter, we intend to report on the present research awareness within language studies. As linguistics today cannot escape the interdisciplinary communication, and the borders of linguistics as a discipline get more and more blurred, the researchers presented here are not only language specialists, but cognitivists and holistic researchers working across the borders of traditional disciplines. Such is the scholarly status of the first researcher already mentioned – Gregory Bateson. Bateson was one of the first to explicitly discuss the necessity of introducing holistic thinking into modern natural, social and humanistic sciences. In his last book, shortly before his death, he presented a general framework for his holistic theory of mind.

The mind is a reflection of the external natural world, a map which is not reality but its representation. The mind is an extension of nature (cf. Bateson, 1996: 15; 128).

Peter Mühlhäusler, professor of linguistics at the University of Adelaide, actively and fervently sets in motion the initial ecolinguistic and holistic proposals. He continues the Haugenian pathway, taking up Batesonian holism, and on this general philosophical/methodological framework builds a new ecolinguistics. Mühlhäusler's publications are no longer philosophical generalizations. As another recognized ecolinguist, professor Alwin Fill, observes, 'it is largely due to Peter Mühlhäusler's work, that ecolinguistics, the study of language, ecology and environment, has developed into a full-fledged area of linguistics' (from the back cover of Mühlhäusler's monograph 2003). In his research, Mühlhäusler confronts several fundamentals of traditional linguistics, pinpointing the weaknesses of Saussurean concepts of language arbitrariness, language conventionalizations, neglect for iconicity or for indexicality in language. He proposes new understandings of meaning in language. He questions the traditional model of the communication process, involving message, code, signal and channel. From the ecolinguistic point of view, this model is very technical, described by Mühlhäusler as 'telegraphic'. He also deals with invariant language: 'even the same person speaking to the same addressee on the same topic, at the same time, in the same place employs variable rather than invariant grammar' (2003: 6). Dictionary compilation or language acquisition research are described by Mühlhäusler as narrow and lacking applicability (ibid.).

## Three types of filters through which we see reality

In his re-definitions and re-modeling of the language processes, Mühlhäusler uses one meta-parameter, which is the notion of *linguistic context*. The context becomes his label for what other scholars – linguists, cognitivists, physicists, or mathematicians – refer to as 'flux', 'primary world substance', or 'the energy field' (see also Chapter 1 and 2, where we look closer at these notions). He writes (2003:45):

> (..) language is actively involved in the process of meaning creation as one of the parameters that constructs our perception of realities. (..) The ecolinguistic position is that the relationship between language and the world is a two-way process.

So, as he maintains, language is essential in a person's contact with 'all-that-is'. And, it is via language – to a large extent – that humans build up and navigate their mental representations of the world.

Language and reality contextualize each other. What needs to be pointed out here as well is the plural form Mühlhäusler uses when saying 'our perception of realities'. It implies the existence of many realities; not one conventionalized, ideal reality that traditional linguists recognized. On the one hand, Mühlhäusler's 'realities' refer to every person's individually-determined and situationally-specified concretization of the virtual idea of reality. We have said in this book that a person through the filters of the senses and the mental structures, perceives a unique, illusionary in fact, panorama of the world. There is no one objective world. On the contrary, there are countless subjective realities, as constructed by the senses and minds of individual people being in, and communicating in concrete situations. Mühlhäusler adds language to the collection of the filters used by humans to perceive reality, which are the senses and the mind. He notes:

> (..) human beings are not equipped with the gift of immaculate perception, but can at best see reality through a number of filters. Of these, language is one of the most important ones (..) (2003: 60).

In its realities-shaping function, language is represented here by lexicon, grammar, metaphors, and discourse (cf. ibid.: 60).

If we take into account the existence of the three filters on the way to our contact with reality, it becomes easier to understand the unfolding of common communicative difficulties in interpersonal communication. What is obvious and rational for one mind and one set of filters, may be non-rational from the perspective of the other mind and the other set of filters (cf. Bogusławska-Tafelska et al. 2010: 25).

The plural form Mühlhäusler uses in talking about 'realities' indicates more than one possible context for language; so, the language process is simultaneously:

- a cognitive process,
- a neurophysiological process,
- an interpersonal process,
- a social process,
- a cultural process,
- a historical process.

Ecolinguists can choose but one out of these aspects/levels in their research into language phenomena; however, awareness of the remaining perspectives is essential for the validity of research. It does affect the research outcome.

## The process of learning

Education is another field of research where ecolinguists relate all their proposals to the underlying context – the unifying field of life. Here, it is Peter Mühlhäusler again who inspired a new pathway. His insights turned out to be the starting point of further local research in educational topics. He writes that 'there is no possibility, for instance, of isolating language acquisition from the acquisition of other knowledge' (2003: 8). The basic assumption that the learning process involves a multilayer net of parameters and mechanisms has become the starting-point assumption of research projects into such education topics as standards control strategy at universities, the emotions in linguistic and educational processes, or the psycholinguistic profile of the language learner and teacher (cf. Puppel, 1999; Bogusławska-Tafelska, 2006a; 2006b; 2007; 2009).

## Communicology. Semiosphere. General Mechanism of Linking

Language defined as a closed system of signs was central in synchronic linguistics initiated by Ferdinand de Saussure (cf. de Saussure, 2004, 2007). Traditional structuralist linguistics neglected so much within the grid of parameters and mechanisms of human language and communication that today a return to the theory proposed by Saussure and his followers seems not possible. At this level of awareness of the context for life processes, language being one of them, linguists talk about signs as embedded in a life-filled, pulsating semiotic space known as the semiosphere. Stanisław Puppel, professor of linguistics, pioneer of ecolinguistics in central European language studies and the founder of the Department of Ecocommunication at the Adam Mickiewicz University, Poznań, Poland, in his publications promotes a broad, dynamic view of language and communication. In his wide-context treatment of human language, he writes about the re-emergence of communicology, introduced in 1978 by Joseph A. DeVito, and today revived as fitting into the new, 'eco' linguistic studies (cf. Puppel, 2008: 11-12). Puppel defines communicology as a multifaceted study of human discourse and communicative practices in both internal and external environments (ibid.). After Lotman, Puppel uses the concept of 'semiosphere'. The traditional model of communication can be applied in research if it is immersed in 'the semiosphere' being the semiotic space. Puppel describes the semiosphere using such concepts as 'matrix', or everembracing and ever-present canvas for the entire interactive semiotic potentialities present in Nature (cf. ibid.). Puppel writes (2008: 13):

> Communicology (..) does not show any such restrictions in its approach to the signs and its interest in the universal 'signing canvas/matrix', that is, the semiosphere, appears unperturbed and unconditionally unlimited'.

Puppel points here to a new broad understanding of signs and the system in which they function. The system of signs, in the ecolinguistic paradigm, is as wide, layered and alive as the entire biosphere of life on Earth; with the 'potentialities' it contains.

In a recent paper (2011), Puppel models the process of communication. This proposal goes far beyond the traditional, 'telegraphic' communication model, including sender, receiver, code, message and noise. The model proposed by Puppel (2011) is an ecolinguistic model embracing:

- communication as the cardinal, even defining process of human life; so the process of communication is embedded within the context of other life processes;

- verbal and non-verbal resources are activated in the communication process, which is based on communicative activeness on two complementary levels: internal (intrapersonal) and external (interpersonal and social);
- the communication practices occur in the global network of communicators via the General Mechanism of Linking.

Puppel sees the General Mechanism of Linking as follows (2011):

> Communicatively, the linking mechanism is secured by the symbol-driven language code which has its universal and language-specific characteristics, as well as by the semiotically and semantically relevant non-language codes governing the management and use of non-language recourses, such as gestures, facial expressions, and body language.

The global network of communicators within, which communication is possible, is considered as a foundation of humanness (ibid.).

# Conclusions

This chapter presents ecolinguistics as an emerging paradigm within the contemporary language studies. The paradigm is the result of two methodological and philosophical shifts, the first of them having occured in the 1970's; the second having emerged at the turn of the millennium. In the 1970's, structuralist linguistics entered the dialog with biologists and sociologists. Since then, the idea of languages being immersed in local and global ecosystems has started to permeate linguistic research. Along with advances in cognitive science, where the mysteries of the mind, consciousness, and neurophysiological processes have started to be addressed again, it has become clear that the ecosystem for language is not only to be understood metaphorically, but also quite literally. Language is not only a mental, cognitive process of the abstract mind; it has material aspects to it as well, as it involves all the physiological/biological resources communicators are phylogenetically equipped with. Moreover, the abstract mind – which allows communication to be carried on, and is the immediate generator of language – builds up in itself a mental representation of reality via the constant contact with this reality, which is both mental and physical. In other words, the mental maps of the world which people have in their minds, and which enable them to navigate life and themselves, are created and updated as a result of the interaction with all-that-is inside their organisms and outside of them. This substance, of various qualities, from very dense and tangible to a mental/cognitive subtle non-matter, becomes an ecosystem for human language on a very direct, material plane, as well as on the subtle non-material plane.

Puppel sees contemporary ecolinguistics as having two pathways: ecolinguistics based on the narrowly understood ecological preservational theory and practices; here, ecolinguistics considers such topics as language contact, language domination, ethinc languages, language death, social and geopolitical aspects of natural languages, etc.; and ecolinguistics based on the deep ecological presumption, that language and communication participate in the preservation of the human species (Puppel – personal communication with the author).

Steffensen in his conclusions aptly notes that 'we leave a Baconian-Cartesian-Newtonian era, and an alternative to its materialistic and monocasual way of thinking is growing forward' (cit. http://southerndenmark.academia.edu/SuneSteffesen/Papers/1055129/Language_ecology_and_society_an_introduction_to_Dialectical_Lingu). Steffensen adds that this new pathway forms both within universities and outside of them.

> At the universities, the scientific enterprise of the 21st century is intimately connected with the abandonment of Cartesian rationality, and this is, for example, anticipated by quantum physics, transpersonal psychology, integral medicine and integral field theories' (cit. Steffensen, ibid.)

Besides being a paradigm, in its broad sense, ecolinguistics also functions as a model, to be specific as a models' model, in contemporary language studies. It allows in all 'local' linguistic research outcome, locates it spatially, within the multidimensional grid of mechanisms, processes and parameters which, as modern science has been proposing, describe Pribram's 'exformation' of the underlying, primary world substance (see Chapter 2). Ecolinguistics arranges, re-defines and contextualises isolated linguistic enquires into selected aspects of the language process. This meta-scientific function of ecolinguistics is vital, because many more insights are found in the language process when divided and secluded linguistic models start supporting and supplementing each other. In the eye of an ecolinguist, contemporary linguistics in general, regardless of the paradigm orientation, does enquire into the same basic unfolding of reality. But it requires an ecolinguistic set of conceptual-terminological tools (and, the conscious distance to them at the same time), to fill the gaps between next more specific or fairly general linguistic models, proposed in the specialist literature today.

# 4 New perception on *mind, meaning* and *cognitive processes*

## Introduction

Linguistic research, in its both traditional/classical and contemporary pathways, puts the notion of linguistic or conversational *meaning* in focus. Modern language studies add the concept of *the human mind* as the generator of language, hence the mind is the co-creator of linguistic meaning. All definitions of, and insights into, the mind and its processing naturally fall within the paradigm active at a given moment of research. Thus, so far in cognitive science it has been the atomistic view of the nature of the world which has provided unfolding for cognitive and linguistic research into the human mind, brain, and language. In this part of the book we intend to trace the shift currently taking place in the methodology of sciences, and in the methodology of linguistics in particular, which is a consequence (or the power force?) of the paradigmatic shift all science is now taking towards the field model of the world.

## Traditional conceptualizations of the human mind

The human mind has been conceptualized in 20th century language studies in the following ways (cf. Bogusławska-Tafelska, 2006):

*Table 4.1. Selected perspectives on the human mind, as grounded in the traditional, atomistic model of the world*

| definitions and metaphors of the mind | consequent assumptions about the structure and functions of the mind |
|---|---|
| 1. the mind is a container | a. it is 'a black box', impenetrable by scientific work, according to behaviourism; and penetrable by science, according to the cognitive stance which in the 1960's gradually took over as inspired by Noam Chomsky (cf. Puppel, 2006); |
| | b. a natural, physical-like organ (Chomsky, 2000); |
| | c. a theoretical, transformational-generative system composed of units and modules (cf. Mandler, 1984; Akmajian et al., 1990; Jackendoff, 1999; Aitchison, 1998; Puppel, 1996); |
| | d. a cognitive network which stores procedural knowledge and declarative knowledge (cf. Lamb, 1998); |
| | e. an engine of thought (Cummins and Cummins, 2000); |
| | f. the mind is a computer (Searle, 2000); |
| | g. the mind is a mental map (Korzybski in Bateson, 1996); |
| | h. a system built up of concepts, scripts, schemas and frames, which, when activated, form a person's awareness; |
| | i. a site of the declarative memory and the procedural memory; |
| 2. the mind is a process(or) | a. the mind is equated with the repertoire of cognitive processes such as: concept formation and activation, problem-solving, decision-making, message reception, message production, bottom-up processes, top-down processes, strategic behavior and others; |
| | b. a life-long process; |
| | c. a self-correcting and self-instructing system; |

This synthesis of definitions of the mind is not intended to show the timing of their emergence; rather, we start from more intuitive and less abstract vision of the human mind as a container-resembling dish; then, we proceed to an approach according to which language is a process. This collection of definitions allows one to notice (i) the atomistic and ultimately (ii) deterministic nature of the cognitive base. The former feature results from the ever-present organizing idea, visible in the tabulated definitions of everything being co-built out of something else; that is, the idea of the structural and hierarchical componential organization of elements, structures and, consequent on these, processes. One can notice a somewhat distressful lack of self-sustainability and a lack of individual choice in these systems, where matter and non-matter are existing and functioning within the already fixed, structured machinery. The machinery of the human organism and the machinery of the human mind are already internally designed and specified, no matter how much science has understood of it at a given point in time. Today, when looking at cognitivism and mainstream language studies one realizes that humans are not only prisoners of their genetic code, but also of their mental resources (and the collective mind's resources). In the cognitive models humans do not act. Rather, they are acted upon.

After a moment of detached reflection, one reaches the intuitive realization (intuition being yet another complex intraorganismic mechanism for tapping into the nature of things) that at this stage of scientific consciousness and mass consciousness evolution, it becomes urgent to widen the view and try to build a new world model which would account for self-autonomy and sovereignty of the human-communicator. Communication studies, ecolinguistic pathways in modern linguistics, new physics and new mathematics have already taken to such a task. As was said at the 'Quantum brain dynamics and the humanities: a new perspective for the $21^{st}$ century' conference, the quantum model of the human brain could be in some sense the prototype for a new concept of making science (cf. Globus et al., 2004: 9). The sections below discuss the new models of the mind/brain, and new approaches to several key topics in the studies of human language and communication.

## Human mind/brain as measuring machinery to browse in the world of possibilities

Let us start the discussion of the paradigmatic and methodological shift and its effects in the mind and language studies by referring to Del Giudice who, in his research into the modern field model of the world, juxtaposes as he puts it 'cheap metaphysics' with 'rigorous physics'. He notes the need to firmly state the scientific and formal character of recent shifting proposals, however groundbreaking and counter-rationalistic they are now. Wolf, in a similar fashion, directly admits that recent proposals in research on the mind and consciousness bring conclusions so bizarre that they might be science fiction (http://www.liloumace.com/Dr-Quantum-Fred-Alan-Wolf-PhD-Time-Space-Matter-Quantum_a1790.html). The ascending 'field' paradigm in modern science, based on rigorous physics and mathematics, in many aspects indeed grounds metaphysical and philosophical considerations of the last 100 years or more.

## Superposition in the world of possibilities

Superposition is the first element of the quantum model to be looked into here; we will refer at this point to the research of physicists, mathematicians, and neurocognitivists. A classical quantum experiment involves an electron which, while in the superposition state, can occupy two positions at the same time, retaining its oneness and integrity. Adopting a wider perspective, as Franck puts it,

> superposition is the mode in which quantum theory allows sums of orthogonal states of a system to exist or, rather, sub-exist without being manifest. These superimposed states are what the so-called state vector of the system is composed of' (cit. 2004: 53).

Hence, it is the mode of superposition which allows the hologram of the world to *sub-exist*. The primary substance of the world, in the form of a hologram, stays in the superposition state prior to being displayed. As Franck says, states of systems are real without being manifest. They are sub-present (cf. Franck, 2004: 54). So, quite literally, one is dealing here with a world composed of not yet realised possibilities. According to this vision of reality, any notion, object, or aspect has more than one possible manifestation. Physicists and mathematicians touched here on what great thinkers, philosophers, and ordinary minds intuitively suspected, namely, that the reality we live in allows free will, creative living, and a choice re-emerging for humans with every life situation.

To further explain the nature of the world as existing before any 'measuring' device breaks the superposition state, we will refer to how Plotnitsky explains this mechanism (2004: 42-43). He says that if we have an electron emitted from a source and then measured at some distance from the source on a photographic plate, we can consider two types of results:

1. according to classical physics, the measurement would give the same electron, at the position predicted exactly by the measuring instruments used;
2. by contrast, according to quantum physics, there are a number of alternatives which can become the result of this experiment: we can obtain an electron, nothing, a positron, a photon, an electron-positron pair, or still something else (cf. ibid.).

Penrose (2004: 604) talks about the contemporary physicist model of elementary particles; in this model, the electron can be seen as an oscillation between the levorotatory particle 'zig', and the dextrorotary particle 'zag'. Thus, the electron has a zigzag pattern, where a step forward continually replaces a step backward. The electron is an averaged out quality, that is, it remains in the quantum superposition.

So, in the quantum world the initial state of superposition arranges the primary substance and in itself contains the alternative/complementary notions, qualities and motions. All matter and non-matter remain underdetermined and complete, as if ready to be determined and reduced to a single option. In the section below, we will see what happens later in 'the world of possibilities' and why, eventually, we as humans, decision-makers and communicators, land on the duality-based, material ground, with binary systems, and one-solution problems. We will try to detect why humans do not perceive and do not realize this array of possibilities in their reality.

## The mind/brain used in ex-formation or in-formation of the holographic world of oneness

As Plotnitsky (2004: 43) explains, 'from the quantum perspective, the brain or the body may be seen as a kind of measuring machinery, a conglomerate of measuring instruments, suited and developed for both classical and quantum measurements'. So, let us define the human brain/mind to be a measuring device which, in quantum terminology, reduces the state vector (marking the superposition state), which means the collapse of superposition and selection of particular choices out of an array of possibilities. In simpler terms, the human brain/mind, when entering the plane of life and starting to carry on its mental operations, becomes a measuring mechanism which immediately affects the initial quantum state of superposition the primary substance 'resides' in.

Once the mind/brain enters the plane, choices start to be made, paradigms are chosen, and the whole system loses its quantum 'all embracing' features, becoming a classical world, measurable by classical devices. This line of reasoning is based on the assumption which governs several recent physical models and which has it that 'brain processes responsible for consciousness and thinking are fundamentally quantum rather than classical in nature' (Plotnitsky, 2004: 29). They become part of the classical world, when the mind/brain makes a choice. So, when introducing a quantum model of the world, scholars do not replace the 'old' classical model with it; rather, the two levels, the primary, deep quantum structure of oneness, and the surface structure of classical world, co-exist and co-build reality.

Pribram (2004: 493) talks about a 'strange description' of the world construction, in which the quantum state for a while remains in the form of a wave function, to be contracted immediately into a local form when the measurement is done. This is the quantum leap, which questions our sense-based perceptions.

Fels (2010: 11) discusses one more aspect of the dual, quantum-classical nature of the world substance; namely, he looks at communication processes, starting from molecules, then intercellular communication, communication of the organism with its environment, etc. Fels argues (2010:11):

(..) cells appear as being constrained to relate at least with the environment. When groups of cells are close enough to relate with each other and will do so, one group still establishes a relation with the environment. Thinking of embryology where cells differentiate into interrelated types of cells one might speculate that this increasing number of types (groups) started from simple relational patterns as described here. Interestingly, when we think of a multicellular organism's information processing regarding its integration in its ecosystem, we find sensory cells that relate with an external signal (e.g. smell, optical cue, or temperature) and nervous cells that relate with these sensory cells. At the single cell level, however, signals will be transmitted from molecule to molecule thus reaching the world of quantum mechanics (..). Within the ecosystems there are even more levels of organization (from photons to molecules, cells and organs, organism, group members, and species communities). These levels are assumed to display (i) the same principles (..) and (ii) (..) the same relational patterns.

Fels points to a pulsating mechanism of quantum-classical communicating layers, which – interestingly – themselves constitute both life processes and communicational processes. Thus, we can see here how essential it is at this stage of research, to lead the dialog between scientific disciplines when discussing human language and communication. Linguistics, in its focus on the conventional communicational code and the mental base navigating this code, seems to reduce many aspects of the communication process. In this preliminary study we intend to present the wide transdisciplinary context for all language and communication phenomena.

## What is consciousness?

This multilayer pulsation of quantum and larger-scale classical mechanisms changes our approach to the problem of consciousness, its definition and the role it plays in the world structure. Generally, there are two differing standpoints to this problem that can be found in the writings of neurocognitivists, mathematicians and physicists keen to model the world and man in it. Namely, there are scholars like Plotnitsky (2004: 29), or Vitiello (in Desideri, 2004: 24) who see human consciousness as the product or process of the human mind/brain. This approach makes it difficult to explain how the superposition state of the primary substance becomes broken/reduced to allow all ex-formatted and in-formatted phenomena to occur in the universe, man included. When Pribram analyzes the tasks of the human mind and the human brain, as operating within the quantum model, he describes the mind as an internalized forming of the primary substance (flux), that is its in-formation. While, the quantum brain is an ex-formation, an externalized or concentrated form of flux, a part of the matter (cf. Pribram, 2004: 232). If we need a consciousness as a measuring device to create a man with a mind/brain, what is the nature of this primary, starting consciousness? Pribram asks this question in his research and to deal with it he proposes another, complementary model of consciousness in which it is not consciousness that works as a measuring device but the life-creating quantum leap which gives birth to consciousness.

In Pribram's proposal the question how the primary superposition becomes reduced to allow life forms and processes, remains open.

It seems that what we need at this point of the discussion is 'the grid approach'; again, we seem to be dealing here with multidimensional machinery, in which two levels of consciousness have to be distinguished:

1. consciousness understood, as Pribram puts it (2004: 992), as a real, physical process happening 'out there', in the universe; this process does not need human conscious attention 'to be' or 'to occur'. The primary holographic level of all-that-is would be operated by, and put to pulsation by this deep-level consciousness, not personalized or senior in any way, but rather naturally working within the holographic oneness of everything;
2. consciousness personalized or localized within a living organism; or within an ego-centered human mind; this level of analysis would see consciousness as a smaller-scale sample of the primary life ex-formatting consciousness. This organism-hosted personalized consciousness would act holographically so that in its limited, personalized form the whole of the primary consciousness could be traced.

We find it essential to introduce this stratification of types of consciousness, if the further analyses of the functions of the human mind and cognition are to be specified in generating, modeling and navigating the emerged reality humans live within.

## Cognition in the holographic model of the world and man: thinking, decision-making, problem-solving, evaluation

Johann Summhammer (2011) undertakes the interesting task of modelling the chain of processes of sensing, deciding and acting in living systems, using the example of two butterflies which find each other although initially separated by hundreds of kilometers. The analysis is based on the idea of the butterflies being non-locally correlated with each other, both being in a quantum entanglement. Summhammer comes to the following preliminary conclusions:

1. the scenario in which the separated butterflies remain in the quantum relation of entanglement is the most probable one; in spite of an initial considerable distance apart the insects meet, which can hardly be explained by classical biological or physical models; it is hardly possible that the scent molecules can reach such far-distant objects, as two butterflies can be hundreds of kilometers apart;
2. the entanglement scenario is also the most economical one, in the sense of, first, energy expenditure; second, the evolutionary pressure, which would be pushing the butterfly to develop biological mechanisms to localize the other butterfly. Quantum entanglement makes the process economical and relatively undemanding;
3. the mechanism of entanglement may be naturally present in biological systems; it explains all non-local, otherwise rather surprising relations between living organisms and within living systems (ibid.).

Summhammer also suggests a very interesting perspective on the quantum unfolding of decision-making, problem-solving or evaluation processes. A pair of butterflies in their random movements and spontaneous decisions maintain the task of meeting each other. All their decisions and choices, chaotic as they appear to be, in effect bring a completion, which cannot be regarded as random. The quantum-level entanglement of autonomous, separated objects brings coherent results within seemingly incoherent relation (ibid.).

When we look at the cognitive system of humans and the decisions or other thought operations humans undertake on an everyday basis, very often cognitive actions are commented upon as irrational, strange or random. Research into decision-making in humans reveals such a surface-level inconsistency and thoughtlessness on the part of people who are facing choices. In one of the research projects, the experimenters used think-aloud protocols to test the cognitive mechanisms of decision-making. The conclusions of the research were that often there is no consequentialism in making decisions or selecting options. As the experimenters observed, the subjects tended to make choices not based on analysis of the consequences. They thought little about the cards which they would not choose. Rather, they were deciding or selecting first, and thought about the alternatives afterwards (cf. Evans, 1996). While on the surface level we have randomness, inconsequentialism or irrationality, the deep-level processing may be non-locally entangled, thus very much consistent and logical. Summhammer (2011: 348) also concludes that, surprisingly, coincidental decisions and choices humans make every day may be resulting from quantum correlations, and thus make sense on another level of reference.

Our research (2010) into the communication process also revealed the non-linear, unpredictable, momentary surface level of the flow of communication. Structure-oriented linguistics, immersed in the tradition of the semantic analysis of the wording of a message, would try to determine the truth conditions in the studied messages. Semanticians would look for true vs. false testimonies, etc. On the other hand, contemporary linguistic investigation, similar to the one we undertook in the 2010 research, would not try to determine 'the winners' and 'the losers' of a given communication situation. All inconsistencies and dynamicisms noticed in the communicative situations are accounted to, first, the cognitive systems of the participants; second, to their culturally different backgrounds; last of all, to the dynamic flow of the discourse itself, which happens in a particular environment with concrete biological, social, situational conditions (cf. Bogusławska-Tafelska et al., 2010). Our research team's open-mindedly approaching the issues we were observing was essential. The open, multiperspective methodology of ecolinguistics was our conscious choice. And we pre-determined the dynamic arrangements and re-arrangements within the parameters of the communicative events we were to observe. Prior to the research process proper, we specified the dynamic psycholinguistic profiles of the subjects, including the parameter of personal experience affecting the course of ongoing communication; we profiled the dynamics of the communicative situations that were to happen; we took into consideration emotional processes in communication, particularly, emotional infectiousness as the major group psychology force. Nevertheless, the results we obtained in our research project surprised us and developed yet another scenario, which we did not predict (a detailed report of the study can be found in Bogusławska-Tafelska et al., 2010). New ecolinguistic-communicative mechanisms were brought to the surface; we did not find them reported or identified in earlier studies of a similar type. The deep, quantum level activity of living systems indicates to cognitivists and linguists a new dimension of life processes, occurring underneath cognitive and communicational processes. So, apparent surprises and momentary, spontaneous shifts within the communication situation can happen due to various non-local correlations, of which communicators are not aware, but which will navigate the communicational mechanisms. The entanglement coming from interpersonal, personal/intimate experiences, from family relations, or cultural/societal relations, often totally irrelevant from the perspective of the semantic, pragmatics or even cognitive aspects of a given communication situation, is an undercover mechanism navigating communication. The wording or the grammatical structure of conversational turns, as well as gesture language, are the surface-level reactions of the communicators; very often nonconscious, automatic reactions, which give no information about the actual communicative motivations. When we know more about this process in its multilayer nature, the scholarly insights into the cognitive, psycholinguistic and ecolinguistic aspects of communication will surpass the most advanced linguistic presumptions. Respect for another point of view and a different communicative-ecolinguistic reaction may then be reached in mutual contacts.

In this context, one can look at Bohm's study of the relationship between the mind, mental maps and natural language (cf. Stamenov, 2004). Natural language reflects and communicates conventional thinking and authomatic behaviours often not matching the context. For the communication or any other interaction with reality to occur transparently and fully, the mental representation has to be transparent to itself (cf. ibid.). In this way, the mind will come into contact with all-that-is (cf. Stamenov 2004: 160-161). As Stamenov relates Bohm's ideas: 'the flash of understanding (..) is not a matter of presuppositions, implications, entailments, inferences, deductive or inductive reasoning, etc.' (cit. 2004: 161).

## Non-local relatedness and the cognitive processes of problem-solving, evaluation and decision-making

If we analyse human cognition from the perspective of the new world model, applying generalized quantum theory, we notice that we now can formulate interesting hypotheses to several mind-twisting research problems cognitivism has been incompetent or confused about to date. Many of perplexing phenomena and processes cognitivists have been observing in human cognition, were put down to the imperfect human nature or character traits; if a communicator was saying something in one communicative situation, and in the next communicative context the same person was sending a contradictory message, we would classify such a behavior as lying, maliciousness (coming from fear or other primary emotion), personality disorder, etc. Generally speaking, many cognitive processes in humans were diagnosed to result from the imperfect psychological or biological construction in the phylogenetic or ontogenetic sense. In the present study, though, we assume that, ecologically speaking, all living systems including man are phylogenetically (and ontogenetically) optimally designed and capable of harmonious and creative functioning within the reality they are functioning in; after all, as we propose in this volume, living systems are creative systems capable of co-creation of themselves and the material/ex-formatted reality around them. On a deep level of sub-existence, sub-existing living systems make choices of how they will embed themselves within chosen internal and external environments. What we claim here is that mysteries and perplexities of human cognition can be accounted for logical and well-designed formulas; the task of modern science is to get to know these formulas. So far in science, the approach was different, namely, what was mysterious about man or the world was declared to be faulty or deficient. From the perspective of new sciences, all we observe in humans and reality is perfectly imperfect, as these are the choices and decisions made on a deep, quantum/holographic level of life.

In course of our research on the strategies of monitoring the educational standards at the Polish university (the years 2008-2010), we looked at the process of control strategy based on anonymous questionnaires students were filling to evaluate the academic courses they were completing. We noticed that students as anonymous respondents often were expressing different attitudes in different settings. In one case noticed in the course of the research, the anonymous questionnaires the group of students were filling in front of and for the representative of the Institute, the attitudes and evaluations of one particular academic course were rather mild, sometimes slightly positive, sometimes critical; rather emotional. The academic whose course was evaluated expressed her disappointment that the students in these questionnaires were rather indifferent about the majority of the objectives she had set for herself and her course. The academic said that when she was reading these opinions and evaluations, she had an impression that there was no intellectual, communal and cognitive connection between her and the students-participants of the course. Nevertheless, when we looked at another pile of anonymous questionnaires (slightly different in content and form), filled by the same group of students about the same course, but organized by the academic teacher herself, at the end of the course to have some feedback about the reception of the course, the evaluations and opinions found in these questionnaires differed substantially. This time, as the academic teacher admitted to us, she noticed mutual connection between herself and her students. In this round of anonymous evaluation, the students showed their understanding, involvement and even enthusiasm about the key features and key objectives of the course they were assessing. In both rounds of evaluation (done by the Institute, and the academic herself), the questionnaire filling was anonymous, done after the end of the regular course.

We were not convinced about the idea of the students being two-faced, not sincere or manipulative. However, having no other clue how to approach this phenomenon at that time, we abandoned the very analysis. Today, within the new methodological paradigm, we look at the mechanism noticed at that time again, and now we can formulate the preliminary hypothesis that in non-local relatedness which lies at the basis of the group mind and group psychology, the communicative reaction or behavior of the communicator depends on the person/people with whom the communicator comes into contact. Interaction partners determine a communicator's response and approach to a given problem. In the example of students giving different feedback reactions to the course in different evaluation settings, we can hypothesize that:

1. when filling in the questionnaires for the Institute representatives, the students-respondents entered the state of quantum entanglement with the institutional body preparing the testing. As if the students responded to the intentions, visions, and general atmosphere in which this anonymous questionnaire strategy was prepared and implemented. Fear, some form of frustration, bureaucracy, and the non-conscious intention to 'catch' the academic teacher at his/her mis-conduct – were underlying emotionally and intentionally the control procedure. These attitudes, intentions and approach were reflected in the students' questionnaires;
2. when filling in the questionnaires for the academic teacher about her work with the students during the course, the students-respondents were in tune with the teacher's intentions, educational aims, her good will and involvement into her work. This time, it was these attitudes and values which were commented upon by the respondents. The rapport between the academic teacher and the course participants was, in a sense, re-established again.

These processes have a specific unfolding; the one cognitivists have not taken into consideration so far in cognitive analyses of evaluation, communication, or other cognitive processes. In a way, these processes 'read' the layer of consciousness or the paradigm within which they come into contact. Non-local correlation cannot be studied or observed by traditional methodological tools (cf. Walach and Stillfried, 2011). This is the task for multidisciplinary scientific research to work out the new tools to study mechanisms of this nature.

## The functional strategy to browse in the world of possibilities

Wolf (http://www.liloumace.com/Dr-Quantum-Fred-Alan-Wolf-PhD-Time-Space-Matter-Quantum_a1790.html) and Walach and von Stillfried (2011) talk about 'the principle of complementarity' recognized in the quantum systems. Wolf explains that we cannot observe the momentum of an object and the location of the object at the same time. Location specifies, fully grounds the subexisting object or process. Momentum is when potentialities or subexisting entities are in motion and represent an array of choices. So, momentum and location are two mutually exclusive, but at the same time collectively required, descriptions (ibid.). If we apply this principle in the analysis of human cognitive activity, we can say that we are in the flow if we keep asking questions without answering them; asking the questions opens space for alternatives to fall into our lives. This is cognitive behavior which does not switch on the measuring function of the mind/brain (which would equate with breaking down the world of possibilities and landing within a chosen paradigm/a chosen reality). The action of answering the question fixes the situation. We are no longer in the flow. We are focusing on one potentiality. It may not be the best for us or the most wanted. So, as Wolf perceives it, asking questions without answering them directs the human-communicator towards the hologram of the sub-existing all-that-is. Once the mind attempts to answer, the exformation of one alternative happens. Wolf outlines here a possible mechanism of our mind's creative construction of our life/reality/material world (cf. ibid.). As an illustration, when a person is facing an emotional situation, i.e. the wife is mad at her husband for not remembering her birthday (which quite often happens to couples), she can choose between two alternative communicational strategies to express herself:

1. wife: You don't love me any more. (an affirmative statement)
2. wife: Do you still love me? (a question which actually has the function of an affirmative statement; here the body language, facial expressions, biochemical signals of the organism help the wife to send her emotions through to the husband anyway; however, the grammaticality of the question creates a space of options).

Emotional messages or communicative behavior can be based on both communicative strategies illustrated by examples 1 and 2. Some people prefer questions, some prefer statements when expressing their emotional-judgmental states of mind. However, according to the deep level of quantum field processing, the affirmative statement fixes the situation; it equates the decision a communicator makes (in most cases non-consciously) with how the situation will be perceived, and thus experienced. Out of the array of subexisting potential scenarios of what the husband actually did or does, the one in which he no longer loves his wife is selected.

Conversely, according to Wolf, a conscious communicator, while in stress or emotionally involved, is asking questions rather than formulating affirmative statements. So, the wife's reaction in example 2 still retains its function of expressing/verbalizing her anger and the accompanying emotions, but the potential reasons why the husband did not remember her birthday, are all still present. However mad she may be, generally it can be assumed that she does not want the no-love scenario to be the one realised. She is mad, but she wants her husband's affection. In order to avoid self-sabotage, as a conscious communicator she should open the world of possibilities and use proper grammar of her speech which does not fix the situation in the unwanted way.

Nonetheless, in the quantum theory, measurement done using the brain/mind means that one of the potentials, or one of potential paradigms is chosen, and turns into the actual state, measurable by classical science (cf. Franck, 2004: 54). So, a vital element of this mind/brain model is its two-system nature. As Jibu and Yasue write, the human brain is a mixed system composed of 'the macroscopic neuron system and an additional microscopic system' (2004: 268). The former is working 'classically', involving neuronal impulses and electro-chemical processing. The microscopic system of the human brain is 'assumed to be a quantum mechanical many-body system interacting with the macroscopic neuron system' (ibid.: 268). Again, one notices here two different planes of processing, the deep one being much 'deeper' than the cognitive science of the last 20 years has proposed in its models.

When science comes up with a new or revised model of the world, there are usually vital theoretical/intellectual, and applicational reasons for such an action. As regards the theoretical and intellectual reasons, this new model of the mind/brain offers essential insights into how life and humans function. The insights are not possible within the previous model. The applicational reasons, in turn, refer to all further practical/technological/ /expert work which will start when a new theoretical, quantum-classical framework becomes internalised by the scientific community. The two-system model of the brain/mind, which is presented here, brings immeasurable advantages as far as further applications and implications are concerned. In the final chapter of this volume, several applications are pointed out as rising on selected intellectual, scientific and practical advantages within this new research and thought pathway.

## The ecolinguistic model of interpersonal communication

The field model of the world and the quantum model of the brain/mind, together with fresh advancements in other sciences, i.e. biology, zoology and ecology of language and communication, profoundly change our vision of how humans perceive and interact with the surrounding world and with themselves (in intrapersonal communication). Research topics of linguistics and communication studies can be addressed from a different point of departure. These leitmotifs, occurring in linguistics, include perceptions on what meaning is, how interpersonal communication develops, and why humans are not only cognitive/individual beings but also social and cultural beings. These topics can be addressed from a new angle. In effect, new standpoints not only open new thought pathways, but – more practically –

make it possible for us to contact with our own potential, very much present and accessible, but forgotten, neglected, or never noticed. When we know what our choices and resources are, we can activate them. Creativity acquires a totally new value when a conscious communicator enters the quantum/holographic world of possibilities.

The first implication within the new paradigm, which in the forefront of linguistics embodies 'ecology of language and communication', concerns a revision of the communication model. What we intend to focus on now is the communication dyad between the participants. To revise the model, the following observations are articulated here:

1. when the human-communicator enters the communicative situation which in itself is initially a quantum system of sub-existing potentialities, the communicator's mind/brain engages in the basic, primary selection process, it breaks 'the world of possibilities' and chooses a paradigm to further operate within. As a result, superposition is reduced, and a reference platform is determined. We are no longer dealing with sub-existing potentials; rather, we are starting communicational behaviours with a number of selected opening parameters. At this point, ecolinguistic research points to the mechanism a communicator's identity, a communicator 'chooses' at the onset of any communicative event (cf. Bogusławska-Tafelska, 2011: 9-23);

2. linguistic meaning becomes a dynamic quality of on-going discourse, while the signs of the linguistic code which are used in the message exchange, gain momentary values within this particular communicative event. As if these signs being re-built and their meaning re-negotiated for the purpose of any particular communication. The conventional and cognitive features of the signs are taken into account, but do not determine the communicational process. The cognitive unfolding of language (conceptual representations, semantic nets), contrary to popular starting-point assumptions, strengthen the re-creative features of linguistic behaviors and communication. The creativity of communicational behaviours starts when one leaves the level of cognition and enters the level of

the ecology of communication. In the sections below, we develop this line of argumentation;
3. cognitive processes of decision-making, problem-solving or evaluation are subjective and constitute a subjective response from the communicator because (a) the experience-based, ontogenetic-phylogenetic cognitive system generates them; (b) the overall cognitive machinery works since it is embedded within the selected, limited framework which has been chosen out of an array of possibilities. All cognitive behaviours happen within the already chosen paradigm, and are to some extent defined and framed by them.

## A communicator's identity pre-parametrising a communicative event

The human-communicator, when entering a communication situation faces a basic two-way choice: they can start communication from the position of an individual being, an autonomous agent, volitional and comfortably using their neurobiological, cognitive, and emotional tools; or, conversely, can pre-select for themselves the role of a unit of the collective mind, be it a society, a cultural milieu, the community of an institution, etc. As a result of choosing the collective mind as a mental representation, the communicator will be to a large extent limited by agreed upon behavioural and conversational patterns, conventions, stereotypes, generalizations and automatic behavior which make up collective minds (cf. Bogusławska-Tafelska, 2011). Diagram 1 (after Bogusławska-Tafelska, 2011) illustrates the choice a communicator faces, and shows four basic communicational relations they can start at the onset of communication.

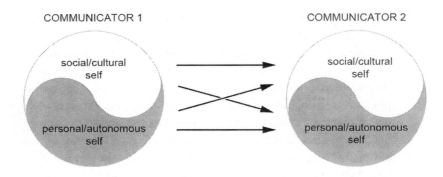

*Diagram 1. Four alternative patterns of communication dyad between Communicator 1 and Communicator 2*

The cognitive model of the mind as constructed out of units and capsules of knowledge can elucidate this mechanism. What the communicator does when choosing an identity prior to a conversational event, is to cognitively preselect a frame of communication (cf. Bogusławska-Tafelska, 2011). In simple terms, frames are mental knowledge structures, collected in the course of previous experiences, stored in the mind, and retrievable for adaptational and conversational purposes. Communication frames are mental models of possible communicational scenarios; a communication frame specifies a person's role and position in a given communication situation. Such a frame specifies the psycholinguistic profile of a communicator for the purpose of a particular communicational event (ibid.). The process of pre-selecting a conversational identity is in most cases non-conscious. At least in those cases when a communicator chooses, often automatically, to represent the collective mind. When aware of the other profile which may be chosen and of the advantages this choice will bring, some communicators are able to consciously select to act from the place of their autonomy.

What is essential from the ecolinguistic point of view is that the choice of the communication identity on the part of the communicator determines (i) their organismic activity while interacting; (ii) consequently, determining the scale of subjection to possible conversation/linguistic manipulation, or conversational failure. What we notice at this point is that the cognitive model of the mental structures (with its references to cognitive frames) is not ultimate in itself; it does not embrace all the organismic processes which start when mental frames are active. We have to leave cognitive linguistics and enter interdisciplinary field of ecolinguistics to embrace all the consequent multilayer processes that happen to a human communicator as a result of choosing the frame of communication, and the communication identity as a result.

## The illusion of language manipulation

What happens when people start communicating from the standpoint of their personal and cognitive autonomy is that this inner selection immediately opens the cognitive, emotional and biochemical programs of the organism, which are phylogenetically present and ontogenetically developed, yet not active most of the time. However, when switched on, these organismic programs navigate the organism to different communicational activeness and then outcome, as compared to organismic programs which start at the 'collective-consciousness mode' (cf. Bogusławska-Tafelska, 2011).

Today, the awareness of these communicational mechanisms, and the awareness of one's own power in the design of the communicative event is quite low. Nonetheless, recently we observe that more and more people are starting to question their subjected communicational roles, as if they are waking up to this expanded understanding of what they are phylogenetically capable of. To illustrate, in the October 2011 issue of the women's magazine *Elle*, in its Polish edition, was an essay entitled *Truth written by the shovel* (English translation – M.B-T.; the original title was: *Prawdy pisane łopatą*). In the text, Jakub Żulczyk, journalist and musician, a man in his late twenties, writes a critical commentary about collective attachment to maxims, which are believed to contain universal truths. Żulczyk says that 'such maxims or dictums are the worst way to understand and structure the world. Especially in crisis situations'. The text contains an elaboration of these ideas, illustrated by examples of Polish maxims which are popularly used and cited; they are, however, in Żulczyk's opinion, entirely useless.

He very consciously observes that 'life is too complicated and multidimensional to be ordered by means of conventional truths and sayings'. What we notice in this text, printed in a popular journal for educated female readership, is that awareness of communicational identity is starting to enter the collective consciousness. The public discourse, which is carried on in such journals as *Elle*, is starting to include such topics as convention in language and behaviours, and the automatic and re-creational effect it brings. Collective consciousness, which is voiced by journalists such as Mr Żulczyk, is waking up to the idea of humans being not only members of the collective mind – conventions and commonly recognized truths as a part of it – but also individual beings whose lives and communicative experiences are and will not be repeated in any other instance, by anybody.

In order to understand what exactly happens in the organism and in interpersonal contact, when we make a decision about our conversational identity at the beginning of any communication, we have to discuss two related topics: the mechanism of micro expressions; and the processes of electro-chemical signaling. The next sections focus on these topics.

## Micro expressions in communication

We left the quantum level of systems and processes to enter 'the classical world' of biochemical, organismic activeness of living matter. This is the level of analysis where macroscale quantum choices have been made, and now the 'ex-formatted' part of the field is starting to live its life in the dense, biological layer. To begin with, one has to bear in mind that communication in the living matter of planet Earth occurs not only by means of vocalized language, or symbol-driven language code. As Puppel notes (2011), communication is a connection which also happens via 'semiotically and semantically relevant non-language codes governing the management and use of non-language resources, such as gestures, facial expressions, and body postures'. Micro expressions are the first category of the organismic mechanisms to enable non-linguistic communication.

Micro expressions are cognitive-physiological programs of the organism which enable purely emotional information to be sent to the interlocutor. They may be described as 'slots' in the physiological and cognitive forms or roles which interlocutors adopt in a communication. Paul Ekman defines them to be fast facial movements which last less than one-fifth of a second and which form a particular 'leakage' providing an information a person does not want or intend to pass on (cf. Ekman, 2003). In a conversation, when two persons start communicating their rationalized messages, the parallel mode of communication opens in which genuine, non-rationalized communicative intentions and emotions are being sent as well. The two parallel channels of communication may agree and send coherent messages, or conversely, may sent discordant messages. Every communicative behavior involves such double layer processing.

Emotions, being cognitive processes, transmit pure, undisturbed information about the sender's intentions and the state of the organism. The functions of cognitive emotions in ego-centric protection are outlined as follows (Bogusławska-Tafelska, 2011, after: Ekman and Davidson, 1994; Maruszewski and Ścigała, 1998; Ohme, 2007):

1. Emotions are always switched on. Cognitively, we function using mental models built with the help of emotions. All that we do mentally is supported and enabled by the emotional mechanisms.
2. The emotional system can be neutral, or can accelerate positive or negative emotions, depending on the ego-centric assessment the cognitive system carries out, an assessment which directs the emotional programs towards some action, non-action being one of the options.
3. Bioneurally speaking (after Ohme, 2007), sensually received information goes for deciphering to the thalamus (nonconsciously) and then to the amygdale for emotional processing (nonconscious processing progresses); later, it goes to bioneural regions which process signals within or beyond the scope of consciousness. In other words, the incoming stimulation, the one entering a person's mind, is initially verified by the nonconscious emotional neurocognitive regions.
4. Emotions judge the stimulation with reference to our well-being.

5. Emotions elicited in the communicators inform each other about each other.
6. Emotions provide feedback information – while we do not always realize what is going on in our minds, our emotional reactions give us information about ourselves. What is essential, is that emotions give us feedback information about our implicit knowledge, to which we have no direct access.
7. Emotions, via micro expressions, enable the genuine intentions of the communicator to be passed on.
8. Emotions enable adaptation.
9. Emotions may be dysfunctional if they are out of sync with the rest of the system; when either the external or internal ecosystem has lost balance.
10. Emotions make it impossible to be objective, as they co-participate in building the communicator's mental map of the world. The cognitive system is not efficient enough, nor is it emotionless (ego-independent), to enable neutral behavior or assessment.

In interpersonal communication, it is emotions which always communicate the genuine intentions of the communicator, via micro expressions which enable emotional information to pour out. This is possible because of emotional programs operating beyond our awareness. Humans are not aware of the subtleties of the processing of the emotional minds, thus, they cannot steer these processes. Although communicators can manipulate their words or even some aspects of their body language, they are nonetheless not able to steer the micro expressions or the biochemical signals they are sending out and which carry the information about their genuine communicative intentions. Thus, quite literally, humans are equipped by nature with mechanisms and cognitive-emotional-physiological programs which, on the one hand, send out the truth about them; and, on the other, enable them to read the truth in others with whom they come into contact (cf. Bogusławska-Tafelska, 2011).

Signaling by touch is another channel of implicit communication which happens between interlocutors. Gallace and Spence notice that 'investigating the possibility of implicit processing interpersonal tactile sensations represents another important topic awaiting further research' (2010: 248). Today researchers know that by means of interpersonal touch, we can communicate at least six different emotional states: anger, fear, disgust, love, gratitude and sympathy (cf. ibid.). It is vital to remember that the skin and its receptors constitute both the oldest and the largest of human sense organs (cf. ibid.). Hence, when we communicate with another person, the wording of the message is very often directed, and rationally processed, and is thus susceptible to manipulative actions; the parallel channels of non-linguistic, non-conscious communication, however, cannot be rationalized, and pre-modelled.

## Emotional expression and cross-paradigmatic phenomena in living systems

The 20$^{th}$ century social and cognitive sciences were immersed in one paradigm, the one which has been chosen for living by the majority of humans across cultures. This paradigm is based on the rule of oppositions, dualisms, binary values governing the world and man's life. The cognitive system (of an individual or a particular community) is the ultimate point of reference and the ultimate governor of the body (whether individual or collective). Different cultures have had their specific expressions of this paradigm. In other words, in different cultural settings people have their local philosophies of life, historiosophically and geopolitically shaped; nevertheless, all these local patterns have had some universals at the basis. The Newtonian, mechanistic model of reality symbolizes this view on life and world.

Hence, when turning to the post-Newtonian model and quantum/holographic/field visions of reality, we are not inclined to reject the findings of psychologists, cognitivists, sociologists, linguists, biologists, and others. Mainstream 20$^{th}$ century research was focused on this particular realization of the holographic potential. The ascending field model of world and life, and the consequent new sciences which have started to emerge at the turn of the new millennium, allow other, novel and unthought so far notions, mechanisms, and assumptions to be taken into account. So, we claim here that we can only talk about universals and patterns if we reduce our point of view to one particular paradigm. Universals do not exist when people cross paradigms and live their lives as creative, conscious beings and communicators.

What is astonishing, new sciences make it possible to observe and start studying parallel possible paradigms which can be realized on Earth, and which occasionally can be noticed today. We can observe small children, as they are good examples of living systems that switch in-between paradigms, and 'create' their reality more spontaneously and eagerly from the wide hologram of all-that-is.

To illustrate our point, we can look at the 20$^{th}$ century research on emotional expression. Paul Ekman has been a leading scholar in this field. As he admits in his book *Emotions revealed* (2003), throughout forty years of his research he was studying emotional expression in divergent research settings; he was observing psychiatric patients, 'normal' individuals, adults, children, people of divergent cultural and social backgrounds. Ekman assumes in his research that there are universal emotions, that are experienced by all humans. The scholar also talks about the universality of facial expressions. As we understand this assumption, universal emotional processes and expressions would have to be inborn and handed down from generation to generation biologically through the genetic code, as universality for Ekman and other psychologists studying emotions as cross-cultural mechanisms, means that irrespective of the culture humans experience these universal emotions in similar or identical ways.

We would like to look at Ekman's proposal from yet another perspective. To do it, we will start from presenting here an examination we did in one of the public kindergartens in the Polish village D. located at the outskirts of Olsztyn. The group of five-year-old children during a regular educational play at the kindergarten, drew three posters with three different emotional expressions: sadness, joy and anger (these emotions are on the Ekman's list of universal emotions). Each child was to draw one face, expressing respectively sadness on the sadness poster, joy on the joy poster, or anger on the anger poster. Then, the posters were exhibited in the kindergarten hall, on the board, where the children's creative outcome is usually exhibited for the parents and guests to see.

# New perception on mind, meaning and cognitive processes 79

*Photo 1.*   The 'anger' poster

*Photo 2.*   The 'sadness' poster

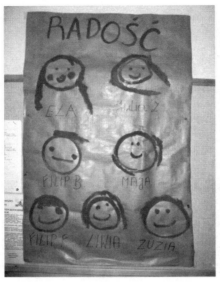

*Photo 3.    The 'joy' poster*

On the poster with anger-expressing faces, one drawing done by the boy named Kostek expresses anger in a non-conventional, surprising way; the face is smiling and in general is received very positively while looking at it. The rest of the faces the children drew are all somehow screwed up indicating angry attitude, thus exhibiting a conventionalized anger expression.

On the second poster, with sadness, again the conventional expressions drew by the children are accompanied by one face which breaks the code; Julia drew a face which does not exhibit sad feelings; the face is smiling, with no sign of distress of any kind.

On the third poster, with joy expressions, one of the drawings also contains an original conceptualization; the boy Filip drew a joyful face which in fact seems to communicate indifference and calmness rather than joy; the joy here is as if present inside the person, not explicit or exalted, as conventionally expressed and conceptualized.

The short enquiry we did with the kindergarten teachers about the three kids who drew unconventional expressions of the emotions, revealed that these five-year-olds are typical, healthy kids. Kostek is perceived to be a very lively, sometimes trouble-making boy.

Genetics, the science of genes, focuses on patterns of inheritance from parents to the offspring and the mechanisms enabling this; a gene is a molecule composed of a chain of four different types of nucleotides – the sequence of these nucleotides equates the genetic information (cf. Wikipedia). New genetics, represented by Bruce Lipton (PhD holder in cell biology) opposes this apparent determinism and says that humans have the genetic profiles but the genes do not activate on their own; rather, humans can switch on or off particular genes through their reaction to the environment. In simpler words, the perceptions and reactions to the external reality decide which genes will be activated within a person's genetic repertoire. So, if we make an assumption that emotional universals are biologically/genetically driven, still we arrive at the plane of the environment or culture a person lives within. Ekman is and is not right at the same time in his claims about 'universal', inherited basic emotional competences.

Culture – as we assume here – is the collective mental representation, the group mind, so to say. If we apply the generalized quantum theory to this analysis, we can assume that cultural influences are possible due to non-local correlations or the mechanism of entanglement. Seemingly separated objects are related to one another, and their behavior is somehow related, through quantum-level correlations. So, whether genetically or culturally steered, humans get involved in quantum-level relationships and mechanisms. And this is the level on which we can enter the holographic field with all possibilities and all life scenarios. Through the quantum level on which the living system operates (the human organism as well), we get into contact and reach the world of possibilities. Small children, not socialized fully yet, more frequently than adults make turns towards other paradigms, than the one 'socially' or 'usually' picked up by the majority of the population. So, to make this reflection less abstract and easier to follow, Kostek, Filip and Julia are not 'unsocialised' or 'incompetent emotionally' as a traditionally trained psychologist could pre-suppose. These children picked up another paradigm of thinking, reacting, perceiving and living the life, unlike the rest of the children.

Present-day psycholinguistic and cognitive research occasionally notices and gives account of this multi-paradigmatic reality we live in. The cognitive and psycholinguistic phenomenon of divergent thinking is a good example here. Dennis Child in his book *Psychology and the teacher* (2007) discusses a dichotomy of cognitive mechanisms in humans, which are convergent thinking and the mentioned divergent thinking. Convergent thinking, in short, refers to binary thinking where the person presupposes in his/her cognitive process two potentialities: good and bad result. Convergers are people who prefer thinking and acting cognitively in the well-mastered lines of thought or action. They will respond to a cognitive task – a problem to solve or a test to take – only if they are certain about their competence /answer in the matter. If not familiar with the right solution – convergers will remain silent. Child says that convergers optain higher scores on IQ tests, and at the same time relatively low scores on divergent thinking tests requiring creativity (cf. Child, 2007: 335–336). Convergers are re-creative while divergers are creative ones. People with preference to divergent thinking are always departing from binary, right-wrong, type of cognitive strategy. They can see and consider more options and more solutions to cognitive tasks, than just the basic right-wrong answer. As they are observed to act, they locate the numerous possibilities they consider in different contexts that can accompany a given problem to deal with cognitively (cf. ibid.).

If further research confirms these very preliminary, sketchy models we present here, the practical advantages of this knowledge are essential, indeed. One can just imagine how much people's lives and their reality would improve if they knew how to enter the world of possibilities and pick up the thinking and acting paradigm which suits better their expectations, intentions, dreams and well-being. This particular metacognitive awareness which enables mediating the paradigms would improve the intrapersonal and interpersonal communication. Judgments based on different perceptions and paradigms the interlocutors often have, would disappear. Again, children are the best example here. Small children are usually positive regardless of the circumstances; children are willing to smile and play in the most harsh circumstances, and when asked how they can do it, they always point to positive, hope-giving aspects of reality, which often are not perceivable by the adult standing beside. If not taught to do so, children do not evaluate reality using the binary scale of good-bad, acceptable-unacceptable. In other paradigms, outside the traditional Newtonian world, the axis of potentialities and choices is not limited to two polarities/choices out of which one is already frown upon (the wrong or bad one). Quite literally, children live in a different, children's world. This is the paradigm they choose out of the world of possibilities in the universal hologram of all-that-is.

## Biochemical signals in intraspecies and interspecies communication

Chemical signaling through body odours and pheromones conveys information between members of the same species, or to members of another species (heterospecific). Reports from research on the mouse and hamster (cf. Samuelsen and Meredith, 2009) indicate that the medial amygdale processes the chemosensory information and decides on its biological and communicative relevance. So, processing of biochemical stimulation is emotional/nonconscious. The vomeronasal organ is pointed out as the sensory organ which receives the signals and sends them to the amygdale. The accessory olfactory bulb and the trigeminal system also participate in this process of reception (ibid.). The following complex behaviours are communicated through the chemosensory channel: status, relatedness, social rank, territorial ownership, reproductive status, and the mental-emotional-physiological state of the sender's organism (cf. Samuelsen and Meredith, 2009: 1468; Rantala et al., 2006; Stockhorst and Pietrowsky, 2004).

Giaquinto and Volpato (1997: 1333) categorise the extraception sensory modalities for receiving signals in fish as follows:

- photoreception
- mechanoreception
- chemoreception
- thermoreception
- electroreception
- magnetoreception.

We can notice how many communication channels are activated in the living organism apart from the traditionally recognized visual-auditory channel. There is no reason to doubt that humans possess these modalities in their phylogenetic package. To illustrate the point, we can look at electroreception and magnetoreception in humans, which are modalities actively co-building the context for communicational processes. The years 2011 and 2012 have seen unusually strong electromagnetic activity of the sun. Solar winds caused by coronal holes and coronal mass ejections on the sun surface have been causing huge geomagnetic storms, which reach the Earth's magnetic field, modifying it and affecting the global ecosystem. Significant disturbances which are caused by solar activity have an impact on moods, psychological/emotional behaviours and the health in general of humans worldwide. Research which will evaluate precisely the range and effects of these transformations is a matter for the near future. Nevertheless, on a daily basis, we can feel these disturbing electromagnetic shifts. We notice intensified geological activity of Earth, changes in the animal world, and increased stress and depression rates in the human population. Public discourse in the media today more then ever concentrates on these issues, which is an indicator of apparent processes.

Hence, in order to model the process of communication it is vital to take into account the impact of the biochemical and electromagnetic environment; it is essential to accept the imperative of multidisciplinary cooperation in determining what affects the communicator prior to, during and after communication. Traditional linguistics fails to explain and improve typical communicational and psycholinguistic problems humans face interpersonally, intrapersonally and in the public communication, the reason for this low effectiveness being the narrow context of traditional linguistic research which omits a grid of multilayer parameters – like the electromagnetic impact of the sun activity – affecting the communicator's organismic functions (language functions naturally included).

Coming back to biochemical signals which build up a pathway complementary to the linguistic pathway of communication, Chen and Haviland-Jones (1999) report on the immediate effects of airborne chemicals on human moods. Their research shows that mood changes are not correlated with the observer's perception of odor qualities. In other words, odors perceived as unpleasant were as likely to incite a depressive mood as were pleasant odors (ibid.). We forget that, as Pause says, from the phylogenetic point of view, hormone-like substances are the most primary substances responsible for intraorganismic and interorganismic communication (cf. 2004). Chen and Haviland-Jones point out in their research that the body odors carry social and biological information; and differentially affect the moods of odor recipients (1999: 250).

Chemosignalling which affects cognitive functions in humans has been the subject of another research project. Haegler et al. (2010) studied the impact of anxiety-related chemosignals on the willingness to take a risk. The cognitive processes of risk-taking, decision-making and anxiety-related behaviors were observed. The research conclusions indicate that 'chemosensory anxiety signals are communicated between humans thereby increasing participants' risk taking behaviours' (cit. 2010: 3901). When received through the chemosensory channel, anxiety can incite in the receiving person the physiological responses like increased sweating, increased heart rate or muscular tension; and behavioural/emotional responses, like the fight, flight or freeze reaction (ibid.). They write:

> '(..) our findings suggest that anxiety in humans can be communicated through chemical senses. The present study confirms previous findings, which showed that chemosensory anxiety signals have effects on cognitive performance, physiological response, and emotion perception (..) (2010: 3907).

So, irrespective of the optimistic or carefree words or linguistic style the communicator may decide to use, the interlocutor will, nonetheless, receive biochemically the genuine information about the communicator's emotional state which, in turn, reflects the communicator's inner psycholinguistic profile in this particular message exchange (we remember that a person's psycholinguistic profile is a dynamic notion as well; it has the attributes of a process rather than a rigid cognitive/emotional - neurobiological structure). Thus, the wording of the message, manipulated or simply rationalized for some reason always goes together with non-linguistically sent information which escapes any attempts to be internally controlled. Naturally the question emerges whether humans are aware enough to apprehend these non-verbal messages. The present study delineates a new pathway in modern linguistics and communication studies, in which a model of communication contains a multilayer grid of processes and mechanisms. Such a wide theoretical view can then be used in applicational tasks, and thus disillusion several communicational assumptions humans have today, and which block their communicational abilities.

Experimental work on behavioural pheromones in humans is considerably intensive today. Pheromones are chemical substances produced by the organism for interorganismic communication. Contrary to signaling by means of odors, pheromone messages cannot be perceived consciously, as they cannot be smelled. Nevertheless these information messengers have an impact on both the autonomic nervous system and on the psychological state of the receiver (cf. Jacob, Hayreh and McClintock, 2001). Human steroids such as androstadienone and estratetraenol have been experimented on to check their role in interpersonal communication, in behavioral navigation and in the social effects of these mechanisms. These steroids were extracted from the skin of human subjects in the research (ibid.).

Pause (2004) reports on studies on androstenone exposure and its communicational and behavioral effects on recipients. The scholar concludes that the research has to be continued, as much is to be learned about the details of this communicational mechanism. Many questions are yet to be answered, however, it is beyond any doubt that humans do use this modality, parallel to a traditional linguistic modality, in message sending and message receiving processes.

Further important evidence comes from the psychological and neurocognitive studies on prosopagnosia, which is a deficit in face recognition due to damage to brain regions reading this type of information. Ohme (2003) refers to the studies of Bauer (1984) and Adolphs, Tranel and Damasio (1998), the first of them providing observations of a man with the inability to recognize faces, who nevertheless displayed the bodily symptoms of recognition. In other words, sight was not the only channel through which the information about the face could reach awareness; the man was receiving the information through some parallel channel; it was experimentally documented that his vegetative system knew whom he was looking at. Adolphs, Tranel and Damasio supported this thesis in their research.

Scientists of several disciplines which focus on man and communication are convinced that interpersonal message exchange is done by means of several parallel channels or modalities. Jarymowicz (2007) writes that often we are not fully aware of our own emotional expression; we go red in the face, we start sweating, etc. Body language provides a rich collection of means of communicating messages. Proxemics and body movements are communicating our truths as well.

In research on cell communication and biochemical signaling there is a field of study, *biocybernetics,* which can direct this section back to the energetic, unifying unfolding of reality (see Chater 2). Ji (1997) explains that biocybernetics sees cellular units (gnergons) to be both units of information and units of energy, depending on the measuring device we use. In their function of energy carriers gnergons, probably can communicate with the universal field from which life on Earth has been ex-formatting. In other words, some recent version of the biocybernetic theory will perhaps make it possible to detect the feedback relation between the biological system (gnergons as information units building the biochemical modality in the communication model), and the energetic, sub-existing hologram of everything, which would be communicating with the gnergons functioning as energy units. Therefore, some properly extended version of biocybernetics will become a communication point between the human body or any other living system, and the primary, holographic world of possibilities.

## Meaning as a process

According to cognitive linguistics, the language process embraces cognitive patterns and the processes of the communicators' minds, momentary and constantly rearranging as they are, together with interacting patterns and processes of the collective, ideal cognitive system (culture), and situational, sociopolitical, geographical and environmental co-participating phenomena. This multidimensional, multithematic context co-creates the meaning and outcome of any communicative act. In the ecolinguistic paradigm in contemporary language studies, which continues and further extends the cognitive approach, meaning is not a property of the portion of language; rather, it is a dynamic compromise, the here-and-now effect of a given communication process (cf. Mühlhäusler, 2003: 9; Bogusławska-Tafelska et al., 2010).

Sometimes, in communication, humans accuse each other of irrationality, illogic, or even deceit. Such communicational disturbances and a consequent lack of mutual understanding on the part of the communicators are very often the result of:

1. language being a process generated by communicators' cognitive systems, which are never uniform in content, are constantly rearranging themselves, and which selectively absorb the information available to them – this is the cognitive phenomenon of bounded awareness (cf. Chugh and Bazerman, 2007). The selected input of the stimulation/information is, again, individual; thus, theoretically speaking, each of the communicators can pick up and focus on different aspects of the shared communication context (ibid.). What is rational for one mind, can be non-rational from the perspective of the other mind;
2. from what we have already discussed in this book we can speculate that communicators involved in a communicative situation are each operating within different paradigms, which they chose at the onset of the communication. This variety of paradigms involved may also cause communication disturbances. Hence, in linguistic considerations, we no longer talk in terms of truth conditions or lack of cross-cultural competence. Now both parties are equally ready to participate in the conversation; both parties have their truths;
3. communication scenario and outcome are subject to internal and external/environmental 'ad hoc' shifts and turns – small disturbances which can ultimately change the communication direction altogether; this book provides a view of the overall panorama in which any communication act is embedded;

4. ultimately, cognitive, biological, situational and cultural contexts for any communicational situation form on the basis of the initial paradigm choice, and then are further navigated by the communicative identity on the part of the human-communicator.

The choice of paradigm, which decides how a person will function in the material world, having been the embodiment out of the holographic field, usually happens unnoticed if not followed by a moment of conscious reflection. The reason is that the model of life, contemporarily popular in the western world, is based on the simple, linear scenario most of us adopt automatically: a newborn child has the full phylogenic potential at the start of life; the socialization process then begins, first at home with the caregivers, and then in an intensified adaptational process in the educational settings the young person starts to function in. Socialization and general education today for people of the western world means gradual dis-balancing of the two identities we are discussing – the individual identity which holds self-confidence and individual sovereignty is gradually losing its importance; while the social/cultural/communal self which holds the models of the collective consciousness, starts to steer the persons mental processes, behaviours and life script.

This choice of communicational identity, however neglected by most communicators, can be made at any moment within the communication situation, and sometimes indeed makes itself available in the course of a communicational process. Some communicators do, occasionally, exercise their personal autonomy in communication. Our 2009 research shows that both types of communicational identity are present in our communication situations (cf. Bogusławska-Tafelska, 2009).

James J. Gibson, who has studied vision, information processing, and perceptual processes for 50 years now, points out that stimulation carrying information (which can be received by the organism) does not have the form of discrete inputs, but the form of continuous flow (cf. Gibson, 1986: 58). He writes (1986: 57-58):

> The environment of an observer was said to consist of substances, the medium, and surfaces. Gravity, heat, light, sound, and volatile substances fill the medium. Chemical and mechanical contacts and vibrations impinge on the

observer's body. The observer is immersed as it were in a sea of physical energy. It is a flowing sea, for it changes and undergoes cycles of change, especially of temperature and illumination. The observer, being an organism, exchanges the energy with the environment by respiration, food consumption, and behavior. A very small fraction of this ambient sea of energy constitutes stimulation and provides information. The fraction is small, for only the ambient odor entering the nose is effective for smelling, only the train of air vibrations impinging on the eardrums is effective for hearing, and only the ambient light at the entrance pupil of an eye is effective for vision. But this tiny portion of the sea of energy is crucial for survival, because it contains information for the things at a distance.

The flow of the physical energy Gibson is talking about, for Pribram has the form of a subexisting hologram of all-that-is. Humans-communicators are immersed in this continuum, and design their life realities, being more or less conscious of their creative role, starting from the paradigm they choose to move within. Meaning in communication becomes another derivative of this overall dynamic. Linguistic/communicative meaning becomes a here-and-now outcome of the paradigms of the communicators; the following features of the communicative situation are connected and/or dependent on this initial choice of paradigm:

- environmental dynamic influences: biological, temporal, historical, cultural/social parameters;
- personal and intraorganismic parameters;
- interpersonal relations.

## 'To see is not to see'

Rafał Ohme (2003: 145) in his book on mimic expressions of emotions relates the anecdote told forth by the British philosopher Ernst Gellner, who notes that today we are witnessing the third stage of the process of naturalization of man: first, Nicolas Copernicus put us, proud people, outside the centre of the universe; second, Charles Darwin put us, again, in-between animals. Then, Sigmund Freud said we are not in control of ourselves and are constantly tormented by inner tensions. As if to support this standpoint, today in studies of cognitive emotions, there are scientific efforts to build a model of the regulatory system in humans, which works outside the essential participation of human consciousness. This anecdote puts man in a rather weak and submissive position within the atomistic, linear world model. This model has underlied contemporary scientific activity in the humanities and social sciences. From the perspective adopted by this model, we are limited by our minds, which navigate our actions and responses; and often act beyond our awareness. Moreover, in the lives we live, we are committed to the social/cultural/community milieu (which is a collective mind), and are navigated by the conventions (communicational as well), rituals, and traditions active in that milieu. Indeed, when one reads present-day research results in studies of man and communication processes, one obtains a picture of very autonomy-deprived individuals, rarely capable of spontaneous creativity, in their actions and re-actions.

Nevertheless, modern science digs deeper into the fabric of life. Quantum and holographic models of the nature of life and universe have left the atomistic level and concentrated on subatomic processes and mechanisms. On this level of functioning, we are the expressions or concentrations of the universal field; we are the ex-formation of the universal wholeness, in Pribram's terms.

The new world model stresses the process of *choice* on the part of life systems. When we are born, we start selecting. Very quickly, our minds choose the paradigm in which they start functioning while living on the Earth. As today humans – most often rather automatically – choose the atomistic and dualistic paradigm – their minds switch on and start developing the ego-centric mechanisms which allow adaptation within the world chosen.

Choice seems to underlie the majority or even all activities of living subsystems. Moreover, within a living system, choice becomes available and starts mechanisms on various levels of the organismic structure: the macro structures often, though not always, choose with the assistance of conscious attention; while on the micro level, structures pulsate from the quantum superposition state of choice to 'classical physics' choice outcomes. What Ohme and other psychologists and cognitivists have been studying in the field of cognitive emotions and neuropsychological and social aspects of emotional activity, remains but a narrow fraction of one possibility out of the 'potential realities' which are available to us.

In the ascending field model of the world, humans are integrated into the oneness of the field. Wolf talks about all humans being projections of one consciousness of the field (cf. http://www.liloumace.com/Dr-Quantum-Fred-Alan-Wolf-PhD-Time-Space-Matter-Quantum_a1790.html). It is in this sense that we are naturalized. At the same time, in the functionality of the world substance, we are self-sustained, independent and volitional. We as living organisms ex-formated out of the prime, unifying substance, have the ability to put into existence sub-existing potentialities; it is our forgotten yet actual capability. Reality around is – primarily – underdefined and manifold; ready to be chosen. When we choose to live in the world of rigid boundaries between values, substances and objects, then what we 'see' around us is limited by our field of vision. No longer is the larger picture available for our sensual examination. As Virillo in Prete (2004) notes, 'to see is to be limited in one's field of vision; to see is not to see'.

# 5 Applications and prospects of ecolinguistics as a new linguistics paradigm

## Introduction

This volume is intended to report on recent novel trends within the studies of language and communication. Throughout the chapters and sections, we navigate the reader's attention towards turns which are taking place in the philosophy of science, scientific methodology, and – as time goes by – non-scientific, public awareness. These shifts have been induced by remarkable findings of collaborating natural and humanistic sciences; findings which build a new model of reality (cf. Chapters 1, 2, 3, 4). Contemporary linguistics, naturally, notices these changes. In this chapter, the time comes to talk about applicational potential that new linguistics holds. Here we but delineate the new pathways and mention several prospects for further applicational work. By no means do we aim to write an exhaustive list of applications; this is not possible at this stage of the research, as the full picture of the potential of new linguistics is to emerge along with the further academic research. Nevertheless, we believe that this ecological approach to language and communication (and the mind as the basic measuring tool and antennae) will make it possible to facilitate communication with oneself as well as interpersonal and societal contacts in all communicational settings and niches.

## Mass communication: collective (non)consciousness, culture, society

Walach and von Stillfried suggest that when we relax some restrictions of the quantum theory proper, the axiomatic framework we obtain can then be applied to any system we analyse. Hence, from the perspective of the generalized quantum theory it can be assumed that the mind/brain and the environment are an entangled whole. 'This generalized version is not quantum entanglement but an analogue form of holistic, non-local connectedness of elements within systems (..) (cit. Walach and von Stillfried, 2011: 185). Composing a whole, the mind/brain and its environment give rise to a non-local correlation in time (cf. Walach and von Stillfried, 2011): group mind/collective (non)consciousness, and social and cultural patterning. In other words, collective entanglements do exist and do affect humans in their everyday cognitive/emotional functioning, as well as within the perspective of their life scenarios. Ritualistic practices fix in place these non-local, non-casual links. Walach and von Stillfried give here the examples of family constellations set in place by tradition, i.e. the act of marriage, which becomes in fact more than just a legal settlement or declaration of mutual emotional bond (cf. ibid.). Indeed modern physics provides evidence that humans when acting within social and cultural milieus, and within communicative groups, are drawn to think and act similarly, along collectively chosen, established and continually repeated lines of reasoning and interpretation of reality. In the cognitive and social research, scholars use the terms 'collective mind', 'mass consciousness', or 'group psychology forces'; quantum physicists talk about 'non-local entanglement' between living systems.

It is another issue to reflect on how to exercise one's individual will and choices when involved in such non-local correlations. It is beyond doubt that cultural and social models have their historiosophical value, and often constitute economical cognitive-behavioural shortcut routes in adaptational processes. However, there are many instances of dysfunctional influence along the lines of non-correlational influence. Walach and Stillfried (2011) point to transgenerational patterning, or the correlation present in divorced couples, which causes psychological pain and emotional turmoil. It is essential to consider also group psychology forces and their automatic, non-conscious and re-creative power over the members of the group. Similarly to the individual mind, the collective mind of any social/cultural group is built upon past experiences. In the cognitive-communicational context, what was true cognitively and communicatively a minute ago when the process of communication was carried on, may be and probably is already obsolete, because of the communicative and ecolinguistic parameters continuous pulsation. Group consciousness and social/cultural bonds do contain much expired or dysfunctional stuff because of this basic nature of the non-linear and dynamic reality we live in; and the mind's persistence in catching further portions of experience in a snapshot, to be used for further application. Mental data have applicational potential; however, generally these applications give in effect re-creation, and a limited success in one's self-realization and self-expression.

Being involved in non-local correlations and, as a result, picking up the cognitive stuff embedded in the group mind, does not always serve people well. Research on the ecolinguistic mechanisms of the educational process, and the psycholinguistic profiles of university students and teachers, done in the years 2005-2010 (cf. Bogusławska-Tafelska, 2006 a; 2006 b; 2007; 2009; 2010), enabled us to observe how great is the limiting impact of group mind patterns on individual students, academic teachers, and, in consequence, on the effectiveness of the educational process. The following mechanisms were identified in the course of the research:

1. the psycholinguistic profile of the 'minimal student';
2. university education which attempts to exhibit overall attractiveness, bring constant excitement, and provide immediate applicability in life contexts.

We will here look closer at these two mechanisms, which in our view disrupt the university educational process.

## Group mind and the educational process: educational dyads

Current psycholinguistic proposals define a person's educational process as being founded on (i) his/her self-education; (ii) and on the dyad relation between the person and the expert/teacher (cf. Puppel, 1999; Tudor, 2001; Bogusławska-Tafelska, 2006a, 2006b, 2009, 2010). Hence, ecological study of the educational process embraces both analysis of student aspects and analysis of academic teacher aspects. In our research in the years 2008-2010, we analyzed the psycholinguistic profile dynamics within a pre-selected educational ecosystem, and with reference to the process of standards control procedures applied at the higher education institution under observation. Both parties in this basic educational relation, the university teacher and the university student, are affected by the quality assessment procedure. Both the students and the academics involved in the quality control procedure create the dynamics of the educational ecosystem.

A functionally effective quality control system has to be based on a thorough analysis of the psychodynamics and sociocultural context of the educational ecosystem in which a given control policy is to be introduced. The discussion of the dynamics of language and language environment – external and internal – is motivated by the assumption that it is momentary, local, situationally-driven analyses and interventions that can lead to functionally potent effects. Hence, to start with, the psycholinguistic profile of the student composing the ecosystem under investigation has to be outlined.

In our 2005 study we observed that the educational ecosystem selected for the research, was dominated by the minimal learner profile. The study was conducted on English Philology students; a detailed report of the methods used and results obtained can be found elsewhere (Bogusławska-Tafelska, 2006a, 2006b). We summarize the results here.

The theoretical model of the maximal learner profile – the optimal learner profile – the minimal learner profile was the starting point of the 2005 study of the student profile representation in the chosen ecosystem. In the model built and applied, the cognitive-linguistic, emotional and motivational features of *the minimal learner* are in one or many respects weakened. One of the assumptions was that even one 'minimally' operating cognitive mechanism brings about further negative cognitive, emotional or motivational consequences in a student in an intrapersonal perspective; and also brings about emotional or social consequences in an interpersonal perspective of the group and university. An internally coherent organism which operates within a coherent environment, always regains balance and refocuses itself to function optimally, while the minimal learner's cognitive-emotional-motivational system becomes entangled in further imbalance. So, the difference between the optimal (though not ideal) learner profile and the minimal learner profile comes down to the condition of self-regulatory mechanisms in relation to the ecosystem. In other words, periodic dysfunctions within the organism happen in every organism's dynamics; however, minimalism here is defined as weakened ability, or inability of the organism to regulate itself. Hence, minimal learners need a dynamically designed educational offer within a conscious educational ecosystem to overcome their cognitive-emotional-motivational limitations. An example of the minimal mechanism that we observed was the defective type of instrumental motivation minimal students were developing in their response to the stress and difficulty they were experiencing during the educational process (Bogusławska-Tafelska, 2006). Very generally, the students were focused on marks and diplomas (instrumental motivation), and at the same time were indifferent to the skills and knowledge these diplomas represent (defective type of instrumental motivation).

Low emotional competence, bringing about emotional illiteracy and emotional inner disturbances, together with defective psychosomatic self-regulatory mechanisms seem to be two main obstacles in the educational process of the minimal learner. The intellectual parameters of the student do not count, because loopholes in knowledge can be made up for during the educational process and, as such, do not block education. It is the cognitive-emotional-social noise caused by the minimal profile domination and the consequent turmoil in the ecosystem, which often disturbs or blocks the educational process (Bogusławska-Tafelska, 2010).

The minimal psychodynamic profile of the student affects the relations with other students, with the academic teacher and, ultimately, has a bearing on the triad relation of the student – academic teacher –university:

(1) the most undesirable aspect of the minimal student building communicative/educational relations with other students is the mechanism of 'psychological radiation' referred to as the group mind or group psychology. The research shows that, on the one hand, minimal students often take up the leadership within a group, and on the other hand, it is also minimal students who are very susceptible to external influence and manipulation. Cognitive-emotional noise exercises a strong impact on the cognitive-emotional-behavioral basis of the minimal student. While an optimal student would self-

regulate, the minimal student becomes entangled in defective mechanisms (for a detailed analysis see Bogusławska-Tafelska, 2006a, 2006b, 2007);
(2) the most undesirable consequence of the minimal student building communicative/educational relations with the academic teacher is the defective or blocked communication between the student and the teacher, which in consequence brings the mechanism of the academic teacher frustration, and growing anxiety, and contributes to the minimal psychodynamic profile development on the part of the teacher (Bogusławska-Tafelska, 2007);
(3) an important consequence of the psychodynamic deformations signaled above is the reformulation of the triad relation between the minimal student – the academic teacher – the educational institution; namely, the educational process at the university is no longer based on discussing, negotiating and proposing knowledge, but on negative emotions and growing anxiety. The anonymous portfolios studied in two series of research revealed that half of the questionnaire opinions assessing the courses were emotional, with no reference to course scope or educational advantages/disadvantages. Half of the students-respondents remembered only strong emotions and focused on these in evaluating the courses (Bogusławska-Tafelska, 2010).

## What are the expectations about university education? The collective mind perspective

Today, the collective mind of the western population is preoccupied with two notions: the notion of time/pace, and the idea of sensual attractiveness. Briefly put, people want their lives and all life experiences to be dynamically arranged, based on exciting shifts and turns, so that all that is experienced incites pleasant emotional uplift and excitement. Westerners today rely on *feel-good happiness*, as Reiss names the type of happiness based on our momentary, sensation-based pleasant impressions (cf. Reiss, 2001: 52). The collective consciousness awaits life which would be structured like a cartoon, or an advertisement: quickly changing scenes with colorful and illuminating pictures, and further exciting adventures for the main character bring a general positive reception of the whole.

On the one hand, such an 'attractive' plane of life, glittering with colors and substance, brings constant excitation in the physiology of the organism, and is thus desired and addictive. From the point of view of cognitive functions, such constant external stimulation prolongs the period of external focus of the mind; Maruszewski and Ścigała (1998) discuss the phase of 'epistemic openness' in the cognitive functions of humans. This notion may be applicable here. It is a phase in the cognitive apparatus of humans when the cognitive structures are 'open' and there is an intake of new data into the system. Psychological will or agreement are essential for this phase to start. Epistemic openness in a healthy organism has to be counterbalanced by the 'epistemic closure', which is the mechanism of going inward and processing, ordering etc. of the input gathered in the complementary phase. We can think of epistemic closure as being the phase of the inner work we do to cope with what we have recently experienced. This may be the moment of 'looking at' our conscious and unconscious mental stuff and dealing with it cognitively/emotionally/psychologically (access to the unconscious being possible indirectly). As this coming to terms with unconsciousness and non-ordered input can be uncomfortable and often painful, people avoid this phase in their cognitive lives; consequently, they function as if they were constantly outside themselves, with no contact with their psycho-physiological organisms.

Coming back to the mental model of university education popular today, it seems that the search for attractiveness and cartoon-like plane of life are framing the model. The mental visions of higher education which most people, students, and students-to-be hold, contain this focus on external environment as the source of constant excitation and fun. Thus, they cherish the visions of fun and constantly heightened excitement coming from multifarious, diverse tasks and experiences. Students hope to find the same quality in their university lives, that they are looking in their private lives. In the meantime, passion for knowledge, knowledge search and knowledge negotiating, building the core of the university, come most of the time from laborious, quiet, humble research work. The passion and fun of research pursuits are the result of long-lasting, quiet activity, focused on more and less attractive tasks. Also, most of the greatest intellectual achievements of humans across the ages are reported to happen in the humble quietness of the researcher's focused attention; in non-action, in fact.

In our research, we found several indications of the attitude and expectations students have about modern university. In one instance, one of the students-respondents communicated such an attitude in their anonymous questionnaire evaluating the academic course.

The mutual relation between course requirements and the students' course evaluations finds its reflection in the commentaries found in the anonymous evaluation documents, such as the one below. The chart below with grades given by the student for each aspect of the course, is followed by the additional remark the respondent gave to supplement the evaluation.

| Lp. | The evaluation criteria | Grades :<br>2 – below expected level<br>3 – satisfactory<br>4 – good<br>5 – excellent |
|---|---|---|
| 1. | How do you assess the quality of the classes/lectures? | 3 |
| 2. | Did the academic teacher present the knowledge in an interesting and clear way? | 3 |
| 3. | Was the academic teacher tactful and friendly for students? | 3 |

Additional remarks: Żmudne zajęcia, wymagające. (English translation: M.B-T: Arduous classes, challenging.)

In this example it can be noticed that low grades given by the student – respondent in the questionnaire (3 threes) are followed by a commentary which sheds light as to what motivated such low evaluation. A challenging course and hard and probably monotonous academic work was ultimately low-evaluated by the student.

## Emergent nature of linguistic signs

Human communication is always fully creative and based on emergent in nature linguistic signs. This mechanism lies – paradoxically – at the basis of communicational tensions and failures which regularly occur in human interactions. When brought to the attention of a communicator, this awareness can eliminate most of emotional, interpersonal and intrapersonal tensions connected with communicational problems. The present section delineates this (eco) linguistic awareness essential to (i) become a self-aware and successful communicator; and to (ii) embrace the complete context for the notions of 'sign' and 'meaning', in order to study them academically.

Broadly understood, the study of signs looks at anything that can stand for something else. This contemporary standpoint has been formulated by Umberto Eco (cf. Chandler, 2007: 2). 'In a semiotic sense, signs take the form of words, images, sounds, gestures and objects' (Chandler, 2007: 2). We do not study signs as isolated objects; rather, we observe them within 'sign-systems' (cf. ibid.), referred to as semiotic spaces or semiospheres (cf. Puppel, 2008). Historically, the study of signs in its both Saussurean pathway of 'semiology', and Peircean pathway of 'semiotics', has been close to structuralism in linguistics in the first half of the 20$^{th}$ century (cf. Chandler, 2007: 3-5). Thus, signs were studied as a part of the closed system, the only context for them being some very narrowly perceived scope of the social context.

Ecolinguistics today, when undertaking the study of meaning in communication, looks at signs within a dynamic grid of semiotic space. Puppel in his study (2008: 13) cites Lotman:

> A schema consisting of addresser, addressee and the channel linking them together is not yet a working system. For it to work it has to be 'immersed' in semiotic space. All participants in the communicative act must have some experience of communication, be familiar with semiosis. So, paradoxically, semiotic experience precedes the semiotic act. By analogy with the biosphere we should talk of a semiosphere, which we shall define as the semiotic space necessary for the existence and functioning of languages, not the sum total of different languages; in a sense the semiosphere has a prior existence and is in constant interaction with these languages.

Let us consider now the model which shows the standpoints of generativism and cognitivism in 20th century linguistics. At the linguistic conference *Nowe zjawiska w języku, tekście i komunikacji IV: Metafory i amalgamaty pojęciowe* (held in Olsztyn, 11-12 May 2012) Henryk Kardela discussed the complementary routes of generativism and cognitivism towards the linguistic issue of meaning in language and communication. The model presented by Kardela was based on Saussurean traditional model of the sign being composed of *signifiant* (the signifier), and the *signifié* (the signified). Looking at the model presented by Kardela we could learn that generativists were studying linguistic sign from the bottom-up perspective and were starting their analyses from the material form (the *signifiant*). Conversely, cognitivism in language studies has been applying the top-down perspective, starting the analyses from the (mental) content of the sign (the *signifié*). This presentation of the two-directional dichotomy in the linguistic research nicely shows the problem we intend to address here. Namely, that both linguistic methodologies, the bottom-up methodology of generativism and the top-down methodology of cognitivism, are dealing with the re-creational aspects of human language and communication. First, the older of the two school of generativism was interested in the grammar of language as the generator of language, which was looking at the material, tangible outcome of the process of language defined as we propose here to be one of life processes. Also, the younger school of cognitive linguistics in its manifold pathways, focuses on re-creative aspects of language; the mind which is the ultimate reference in cognitive linguistics, is re-creative in its capability and strategies. The human mind/the mental representation is built out of (i) collective mind material, and (ii) individual life story; thus, in both sources of mental material the mind deals with the past/history. In the meantime, the present cannot be handled optimally by the mind which is always, so to say, embedded in the past. When we receive the information about the present moment, be it in a conversation or any other type of interaction, intraorgnismic or interorgnismic in character, we receive this information as if outside the cognitive map of the mind (we shall not, however, mix mind with brain; human brain may and probably is the coordinating site for this complex set of processes). The present is accessible to us through the past-oriented filter of the mind as cooperating with other organismic modalities such as emotional processing expressed through micro expressions, chemosensory signaling, body movements, or non-local, quantum relations connecting us to the rest of the environment. Even more to it than this, the quantum level of interaction between and inside living systems makes the whole mechanism transpersonal. What if the process of meaning building and meaning sharing within any interpersonal communication was a process going outside the confines of the organisms of individual communicators? It seems essential to consider such a model in which the meaning of any portion of language as emerging within a communicational event is the – essentially momentary – outcome of the negotiation and cooperation between all layers of life mechanisms. If linguistics reduces this fully complex and dynamic mechanism to be but a process of the mental structures re-creating stored previously knowledge under the inspiration coming from the outside, then the notion of meaning is undoubtedly easier to study; at the same time though, it remains unrecognized in its complexity.

As another participant to the aforementioned Olsztyn conference Agnieszka Libura related, communicational metaphors realized in the visual code are stronger in their impact than metaphors realized in the verbal code. This realization cognitive linguists have come to supports our multimodal, grid approach to language and communication. Namely, the process of language as a life process embraces more than just the material form of the language spoken, written or thought. The process of language being a life process embraces all that the communicator says, hears, feels, smells, touches, and is embedded in. As communicators can and usually do live in different paradigms, their minds co-creating the ex-formatted reality of their lives, they usually mean, hear, smell, feel and are connected to different stuff than their interlocutors. The metaphor, or other carriers of message, is potent and well-received by the receiver's organism (is 'strong' as Libura puts it) if it is sent and received by the multilevel living system of different modalities. If, conversely, the metaphor is but a material form formulated by means of one modality, it is a metaphor with a weak impact, if any.

While discussing the multilevel, grid model of sign and the process of meaning emergence, we look at the interesting study of Rossano (2011) who in the research looks at ritualistic and shamanic practices. Rossano says that 'rituals of focused attention created the selective environment from which uniquely human cognition emerged' (2011: 40). According to the scholar, it is ritualistic practices based on well-developed working memory, that through altered/extended consciousness were enabling humans to cultivate their social and environmental relations as well as better health and, in effect, longevity. The scholar says:

> Coolidge and Wynn (Wynn and Coolidge, 2007) have built a compelling case that the emergence of uniquely human cognition resulted from a slight but significant increase in working memory capacity. This increase made anatomically modern humans (AMH) better able to hold information in mind, especially information about behavioral procedures and intended goals (..). *Thus, when confronting cognitive challenges, AMH were better equipped to resist mental sets and other prior habits of thought and behavior. This ability was essential for exploring novel relationships, engaging in cognitive innovation, and ultimately creating and using symbols* (..) (Rossano, 2011: 40) (emphasis – M.B-T.).

Later, we read:

> When our ancestors were engaging in rituals around a campfire, focusing their attention on the flames or chanting a repeated phrase to the incessant rhythm of a pounding drum, they were very likely taxing the very brain areas involved in attention and working memory. Those whose brains were most 'ritually capable' by virtue of increased working memory and attentional control capacity would also have been the ones to reap the greatest health and fitness benefits (2011: 40).

What we can see in Rossano's research (2011) is the following:

1. ritualistic/shamanic/meditative practices are accountable for cognitive advance of the human; it is these cognitive behaviours that were in the past and are today stimulating the evolutionary progression of human cognitive potential;
2. ritualistic/shamanic/meditative processes are accountable for the success in social and environmental interaction;

Applications and prospects of ecolinguistics as a new linguistics paradigm    105

3. biocognitively, these processes operate on working memory, building up the essential state of focused attention;
4. ritualistic/shamanic/meditative practices have been found to be present in the earliest medical healing practices; Rossano writes about the archeological, comparative and anthropological evidence for the universality of ritual healing practices across traditions and cultures;
5. the archeological/historical findings are referring to 'shamanic' practices; today, the respective behaviours are classified as 'ritualistic' or 'meditative'.

Interestingly enough, social, individual and health-related success is connected with working memory, which is involved in processing all incoming stimulation being received by the organism at a given moment. Again, it is the present perspective as accessed by working memory mechanisms, which is considered here. The present perspective, that is the incoming stimuli as processed by working memory, is said to allow life success. It is vital to remember that much of the input that is processed by working memory, never lands in a person's mental representation; rather, it is worked through at the moment and then abandoned. What if similar mechanisms were in charge of communicational success, where the sign emerges within a multilayer grid of interaction and is immediately (one would say 'momentarily') processed by working memory to negotiate communicational meanings. When the communicational situation is completed, the sign fades away. In fact, it may be possible that humans do not rely too much on long-term memory resources for the process of sign emergence and meaning negotiation.

The very mechanism of ritualistic/shamanic/meditative immersing in the 'here and now' experience, which is reported to condition optimal living and life success, may be allowing the direct connection to the universal field of all-that-is. When immersed in the present and with all the organismic modalities open and active, the human communicator may be contacting with the rest of the world substance, the primary consciousness; hence, optimal, full living is realized. Again, this is another thought pathway, building up the frames of new linguistics delineated in this volume, which needs thorough and open-minded experimental documentation. How future (eco) linguists will organize their research seems a challenging task; we believe, though, that this research is the subsequent phase in the studies into man being, acting and communicating in the world.

In conclusion, in this section we would like to propose a new understanding of the emergent nature of the sign; the sign is emergent in the sense of the transpersonal, even universal zone it is co-profiled by. When co-profiled universally and transpersonally, the sign then becomes a building block of genuine creation (and not re-creation). It is profiled 'here and now', within the interacting communicators, their paradigms of life, and the totality of the environment of a particular communicative event.

## Paradigms colliding: autism, ADHD and similar profiles of a new human

When a few years ago we had a pleasure to talk to Professor Tomasz Maruszewski, one of the Polish leading psychologists, he said that as he noticed recently it is every ten years or so that his models profiling human condition and behaviours need revision. Ten years and psychological books lose their up-to-dateness. So, it is essential to note that there are old books which are no longer a source of truths about the human species; and there is old knowledge which can no longer be applied while analyzing humans. According to psychologists like Maruszewski, today the pace is remarkable indeed, with which humans evolve neurocognitively, psychologically, socially and emotionally. As a result, psychological models get obsolete quicker than ever before. CBOS was reporting in 2011 that only 32% of Poles 'were recently reading some book for pleasure' (www.newsweek.pl). The basic reason for people not reading much nowadays can be that people no longer find their truths in books. One can risk an observation that evolutionary shifts are so complex presently, and so frequent, that social and cultural systems seem to lag behind the dynamically changing human consciousness and biology.

The contemporarily descending world model, still dominating:

- collective consciousness
- public discourse/media
- all major social systems worldwide, i.e. the educational system, the health care system, etc.

is based on convention, rigid, automatic, re-creative strategies of thought and behaviour. This is the cognitive-behavioural style inherited from the Newtonian/Galilean model of reality. While being embedded in the Newtonian model, there is no possibility to understand and accept other paradigms and world models; one has to leave the Newtonian world to be able to notice the multilevel sub-existing and existing reality, pulsating with divergent life systems and founded on the basic strategy of 'choice'.

As has been pointed out in the previous section, it is children who are un-socialized and un-programmed enough to enjoy and exhibit the possibility to live according to other paradigms and other world models. Some children are at the forefront of evolutionary mechanism. Today, these children are by the public mainstream systems defined as disordered; in the descending paradigm, the symptoms of behavior which go beyond socially fixed convention, are named autism, Asperger syndrome, ADD or ADHD. Interestingly enough, on numerous internet sites one can come across divagations whether ingenious humans such as Einstein, Newton, (Bill) Gates or Jefferson, to mention but them, were the undiagnosed cases of either of these 'conditions'. Nobody dares, though, in the mainstream public and/or academic discourse, to seriously put to question and debate the whole idea of classifying this collection of symptoms as 'disorders'. One can find lists of the symptoms of the conditions we discuss here in publications devoted specifically to the very topic (see for a quick reference http://www.webmd.com/brain/autism/autism-symptoms). It is vital to notice that the symptoms of the autisitc condition or Asperger syndrome are very uneven, based on blurred methodology of classification, and do not necessarily suggest a problem or a disorder; rather these symptoms may suggest a new evolutionary level of living and communicating.

As concerns behavioural and communicational distractions, fits of anger or other intense emotional outbursts, and the mis-behaviours which new kids perform, may suggest scraping between paradigms. A child feels (with his/her right brain domination) strong discomfort, frustration, or pain when confronting the materialistic, internally incoherent paradigm the adults around live by. Fits may suggest great discomfort of looking into or joining this incoherent, full of polarisms paradigm of the adult world. The new kid looks into or tries to join it when wanting to connect with it and communicate. Literally, such a child experiences the pain of the paradigms 'scraping'.

As a new child is more open to the environmental multilevel reality, he/she suffers from entering the rigid, incoherent (in terms of mutually disharmonious levels of life) reality of the old human (care givers, the old educational system, other old systems in modern societies). We have to remember that the organism of 'a Newtonian human' has many of its mechanisms inactive or underdeveloped; as a result, the semantics and pragmatics of the materialized language are the basis of communication and information extraction. In the meantime, a new child has many more of his/her organismic systems active and processing, hence, he/she is, on the one hand, better informed as for the multilevel aspects of the message put forth (the apparent joke may in fact be a bitter or frustration-driven confession, one would rather cry over not laugh); on the other hand, the child can very easily get overwhelmed and tired from what he/she learns continually just being in the world (cf. Miller, 2008). New therapeutic programs for autistic children and their families, gaining popularity in the US, are based on such a novel perception of autism, Asperger syndrome, or ADHD. These programs point to the evolutionary shift towards new life models, based on totally novel for 'a Newtonian' person choices and life strategies.

Suzy Miller is a therapist working with new children in the mode of, as she puts it, 'energy medicine' (by profession, Miller is a speech-language pathologist) (cf. Miller, 2008). She argues in her book and in her therapeutic work that autistic children do communicate, do feel and exhibit social/interpersonal bonds; and, in fact, they seem to be much deeper involved in these processes than a typical Newtonian-world person. What we need to employ in order to grasp the deepness and multidimensionality of these strategies, is a new awareness of how reality is built and how it is functioning. Miller's interesting account and her successful professional experiences fit well indeed with the new, post-Newtonian world model and the perception on new linguistics as introduced in this volume. We believe that at this stage it is but an onset of a serious debate which is to enter the collective consciousness world-wide and is to be heard in the public and/or academic discourse, soon. New children who personalize the new profile of human, are the catalysts of the paradigmatic change we are foreseeing. Academic research often remains within the closed confines of the academia; what mathematicians, physicists or ecolinguists say does not easily pour into the public awareness. However, new children, through who they are cognitively, biologically, and psychologically and how they live, bring the new awareness to the larger audience of ordinary people across continents, cultures, and social strata.

## Conclusions

Globus notices that 'the idea that we are fundamentally quantum creatures in our functioning is one of the most intriguing ideas of our time' (cit. 2004: 87). This scientific assumption, pioneered in the 1970's by Polish scholar Włodzimierz Sedlak (cf. bioelectronics, quantum basis of the evolution of language, Sedlak in Puppel, 1995), has attracted intensive scholarly attention, and today Wolf, Goswami, Globus et al., and Walach and von Stillfried emphasize in their studies that there is no primary separation between humans, because all matter in the universe is a projection or condensation of the hologram/field unfolding. And the two aspects of reality which are the primary non-matter and its more condensed material part, co-create all life on Earth and in the universe.

At the same time, humans, with their cognitive systems and physiological/biological organisms, have the tools and abilities to design their lives, and choose the parameters of the reality they live in. Bizarre as it may sound, this world model is based on the calculations and measurements of contemporary sciences. Naturally, the research is continuing, missing pieces are being added, and incoherent elements of this model are being smoothed out.

Ecolinguistics is indeed interested in how the non-local entanglement mechanisms work and how to preserve one's sovereignty while in contact with a community we live in or come into contact with. The question is whether we can develop effective strategies enabling social contacts without getting entangled in collective – not our own – perspectives, decisions and patterns of action/reaction. How can we communicate with people around without getting involved in automatic actions and reactions? We do not want to non-consciously be drawn into cognitive-emotional (life) models which, as the research on group psychology and group behavior shows, are always 'minimal' within a given group; in other words, which reduce the intellectual, emotional and behavioral skills of the group members to equate them with the skills of the least intellectually, emotionally and behaviourally potent person who has the leader profile and acts as the 'leader' of the group/community, etc. Walach and von Stillfried (2011) point to ritualistic practices as enabling to set in motion and, probably, dissolve as well non-local entanglement. When we think very generally about the mechanism of rituals, in the wide context of all group communication patterns, their role is to make an individual focus the conscious attention on some aspect of their life and declare their autonomy. This declaration can start the organismic programs we labeled 'the individual/autonomous' mode, in the section on communicative identities humans can adopt. Another strategy which could be of use here is mentioned in the previous sections on asking questions without waiting for the answer; Wolf (Wolf http://www.liloumace.com/Dr-Quantum-Fred-Alan-Wolf-PhD-Time-Space-Matter-Quantum_a1790.html) maintains that through the questions we ask, other possibilities, new, alternative potentials can enter our lives. Entanglement no longer exists. We loosen the fixed composition of the ex-formatted reality.

Secondly, ecolinguistics is interested in how to educate people towards re-establishing contact with their own phylogenic abilities. These pivotal yet still very basic privileges of choice humans have are to be re-established in such abilities, through the educational influence which will be consciousness-raising education.

Ecolinguistics as a new linguistics goes far from the semantic or pragmatic models on how to determine the conversational meaning; or how to map the process of communication. One of the aims of this book was to formulate an alternative view on human language and communication mechanisms. To summarise the new, delineated here methodological proposals, when we want to understand (i) how humans – communicators formulate messages, (ii) how they decode and receive messages, and (iii) what principles or models represent the communication process in simple and complex communicational situations (i.e. when one says 'yes' and actually means 'no'), the ecological approach is the 'grid approach'. That is to say, every action, behavior, or linguistic activity a person undertakes, activates a multidimensional grid of life processes. In this outline study, we were trying to at least identify and pinpoint the varied layers and types of processes that co-build the primary, underdetermined substance and the exformatted life forms, which together form the actual context for language and communication. While linguistic convention, studied thoroughly in the last two hundred years by linguists world-wide, is activated as well, its impact and role are far less important than traditional structuralist or neostructuralist linguists took for granted. For linguistic convention, saying 'yes' is to mean agreement, and saying 'no' is to mean disagreement (to take a simple example); this is the starting point for a linguistic/communicational exchange. While for an ecolinguist, at the very beginning of a conversation, the participants start the process of 'reading' themselves and each other to verify each word, phrase or utterance. They do this along numerous axes simultaneously:

1. they read the message sent and received through a social/conventional/cultural veil, but also read it from their own personal, individualized or even intimate standpoint. Throughout the conversation, every conventional meaning is checked as to whether it is a communicator's personal, inner truth or not. This filtered perspective – be it conscious or not – is always present in a conversation;

2. in addition, the conversation participants send and receive messages by means of non-linguistic channels, i.e. through body language, facial expressions, body odors, biochemical signals. Here convention is absent;
3. all rationalizations the communicators make for various personal or social reasons are simultaneously complemented by the messages their bodies send about the inner, genuine intentions. Micro expressions which communicators produce send this type of information. The co-participants in the conversation receive these signals together with the wording of the message;
4. field of all subexisting and already ex-formatted life builds up the environment for every conversational event. Hence, quantum and classical mechanisms co-build the communicative situation. Such mechanisms as non-local correlations or superposition make every communication event a fully creative process. In this sense, communication becomes a negotiation of parameters, as all the layers of the grid co-participate in the here-and-now process of communication being a dynamic, creative contact of a group of ex-formatted beings in the arena of life.

# References

Aitchison, J. 1998. **The articulate mammal. An introduction to psycholinguistics.** London, New York: Routledge.
Akmajian, A., Demers, R., Farmer, A.K., and R.M. Harnish. 1990. **Linguistics: an introduction to language and communication.** Cambridge/Massachusetts: The MIT Press.
Bateson, G. 1996. **Umysł i przyroda. Jedność konieczna.** Warszawa: Państwowy Instytut Wydawniczy
Blackmore, S. 2007. **Consciousness. An introduction.** London: Hodder and Stoughton
Blackmore, S. 2009. **Conversations on consciousness.** Oxford: Oxford University Press.
Bogusławska-Tafelska, M. 2006a. **Self-education as a strategy of life. The psycholinguistic profile of the Polish student of English.** Toruń: Wydawnictwo Adam Marszałek.
Bogusławska-Tafelska, M. 2006b. 'Defective instrumental motivation as a cognitive rescue mechanism of the minimal learner'. In Puppel, S. (ed.). **Scripta Neophilologica Posnaniensia.** 25-34.
Bogusławska-Tafelska, M. 2007. 'Edukacja studenta minimalnego: praktyczna aplikacja trychotomicznego modelu maksymalno-optymalno-minimalnego'. In **Acta Neophilologica IX.** 81-96.
Bogusławska-Tafelska, M. 2008a. 'Cognitivism in linguistics. Why sciences are to fall into one interdisciplinary paradigm'. In Puppel. S. and M. Bogusławska-Tafelska (eds.). **New Pathways in Linguistics.**
Bogusławska-Tafelska, M. 2008b. 'Kwantowa natura języka. Modele hybrydowe w studiach nad umysłem i językiem'. In Kiklewicz, A. and J. Dębowski (eds.). **Język poza granicami języka.** 197-209.
Bogusławska-Tafelska, M. 2009. 'Quality management and standards control strategy implemented at Polish universities after the Bologna Declaration: a psychodynamic perspective'. In Puppel, S. and M. Bogusławska-Tafelska (eds.). **New Pathways in Linguistics.** 49-72.

Bogusławska-Tafelska, M. 2010. 'Quality management and standards control strategy at the Polish university: an ecolinguistic study'. **Journal of Language and Linguistics**. Baku/Azerbaijan: Progress IPS Company. 47-55.

Bogusławska-Tafelska, M., Świderska, S. and K. Wiśniewska. 2010. 'The ecolinguistics of interpersonal communication'. In **New Pathways in Linguistics**. Puppel, S. and M. Bogusławska-Tafelska (eds.). 21-38.

Bogusławska-Tafelska, M. 2011. 'Manipulation in communication. Or: how ecolinguistics returns a communicator's powers back to him/her'. In Puppel, S. and M. Bogusławska-Tafelska (eds.). **New Pathways in Linguistics**. Olsztyn: KFA. 9-23.

Bogusławska-Tafelska. M. and B. Cierach. 2012. 'Playing with paradigms: Shakespeare's *Hamlet* in the eye of an ecolinguist'. In Koszko, M., Kowalewska, K., Puppel, J. and E. Wąsikiewicz-Firlej (eds.). **Lingua: nervus rerum humanarum**. Poznań: Wydawnictwo Naukowe UAM. 99-110.

Brown, H. D. 1987. **Principles of language learning and teaching**. Englewood Cliffs, New Jersey: Prentice Hall.

Cat, J. 2007. 'The unity of science'. http://plato.stanford.edu/entires/scientific-unity/ DOA: September 2007.

Chandler, D. 2007. **Semiotics. The basics**. London/New York: Routledge.

Cibangu, S. K. 2010. Paradigms, methodologies, and methods. **Library and Information Science Research**, vol. 32. 177-178.

Chen, D. and J. Haviland-Jones. 1999. 'Rapid mood change and human odors'. **Physiology and Behaviour**, vol. 68. 241-250.

Child, D. 2007. **Psychology and the teacher**. New York: Continuum.

Chomsky, N. 2000. **New horizons in the study of language and mind**. Cambridge: Cambridge University Press.

Chugh, D. and M.H. Bazerman. 2007. 'Bounded awareness: what you fail to see can hurt you'. **Mind and society**, 6. 1-18.

Cummins, R. and D.D. Cummins (eds.). 2000. **Minds, brains and computers. The foundations of cognitive science**. Oxford: Blackwell.

Dąbrowska, E. and W. Kubiński (eds.). 2003. **Akwizycja języka w świetle językoznawstwa kognitywnego**. Kraków: TAiWPN Universitas.

Del Giudice, E. 2004. 'The psycho-emotional-physical unity of living organisms as an outcome of quantum physics'. In Globus, G. G., Pribram, K. H. and G. Vitiello (eds.). **Brain and being. At the boundary between science, philosophy, language and arts**. Amsterdam/Philadelphia: John Benjamins. 69-85.

De Saussure, F. 2004. **Szkice z językoznawstwa ogólnego**. Warszawa: Wydawnictwo Akademickie Dialog.

De Saussure, F. 2007. **Kurs językoznawstwa ogólnego**. Warszawa: Wydawnictwo Naukowe PWN.

Desideri, F. 2004. 'The self-transcendence of consciousness towards its models'. In Globus, G. G., Pribram, K. H. and G. Vitiello (eds.). **Brain and being. At the boundary between science, philosophy, language and arts**. Amsterdam/Philadelphia: John Benjamins. 21-28.

Drogosz, A. 2010. 'Existence is life. Metaphors of language that ecolinguistics lives by'. In Puppel, S. and M. Bogusławska-Tafelska (eds.). **New Pathways in Linguistics**. 59-73.

Ekman, P. 2003. **Emotions revealed**. New York: Times Books.

Evans, J. 1996. 'Deciding before you think: relevance and reasoning in the selection task'. **British Journal of Psychology**, No. 2. 223.

Fels, D. 2010. Analogy between quantum and cell relations. **Axiomathes**. Springer: published online 10 June 2011.

Feyerabend, P. 1993. **Against method**. London: Verso.

Filk, T. and H. Römer. 2010. 'Generalised quantum theory: overview and latest developments'. **Axiomathes**. Springer: published online 13 November 2010. 211-220.

Fill, A. and P. Mühlhäusler (eds.). 2001. **The ecolinguistic reader**. London/New York; Continuum.

Finke, P. 2001. 'Identity and manifoldness'. In Fill, A. and P. Mühlhäusler (eds.). **The ecolinguistic reader**. London/New York: Continuum.84-90.

Franck, G. 2004. 'Mental present and the temporal present'. In Globus, G. G., Pribram, K. H. and G. Vitiello (eds.). **Brain and being. At the boundary between science, philosophy, language and arts**. Amsterdam/Philadelphia: John Benjamins. 47-68.

Gallace, A. and Ch. Spence. 2010. 'The science of interpersonal touch: an overview'. **Neuroscience and Biobehavioural Reviews**, vol. 34. 246-259.

Giaquinto, P.C. and G.L. Volpato. 1997. 'Chemical communication, aggression, and conspecific recognition in the Fish Nile Tilapia'. **Physiology and Behaviour**, Vol. 62, no. 6. 1333-1338.

Gibson, J. J. 1986. **The ecological approach to visual perception**. New York: Taylor and Francis Group.

Gleick, J. 1998. **Chaos**. London: Vintage Books.

Globus, G. G., Pribram, K. H. and G. Vitiello (eds.). 2004. **Brain and being. At the boundary between science, philosophy, language and arts**. Amsterdam/Philadelphia: John Benjamins.

Górska, E. 2010. 'Life is music. A case study of a novel metaphor and its use in discourse'. **English Text Construction**, 3:2. 275-293.

Gregory, R. L. (ed.). 2004. **The Oxford companion to the mind**. Oxford: Oxford University Press.

Haegler, K. Zernecke, R. et. al. 2010. 'No fear no risk! Human risk behavior is affected by chemosensory anxiety signals'. **Neuropsychologia**. Vol. 48. 3901- 3908.

Halliday, D., Resnick, R. and J. Walker. 2005/2005. **Podstawy fizyki**. Vol. 1-5. Warszawa: Wydawnictwo Naukowe: PWN.

Halliday, M.A.K. 2001. 'New ways of meaning'. In Fill, A. and P. Mühlhäusler (eds.). **The ecolinguistic reader**. London/New York: Continuum. 175-202.

Haugen, E. 2001. 'The ecology of language'. In Fill, A. and P. Mühlhäusler (eds.). **The ecolinguistic reader**. London/New York: Continuum. 57-66.

Hogan, K. 2010. **The psychology of persuasion**. Gretna: Pelican Publishing Company.

Hutchins, E. 2010. 'Cognitive ecology'. **Topics in Cognitive Science**, 2. 705 -715.

Jackendoff, R. 1999. **Languages of the mind**. Cambridge/Massachusetts: The MIT Press.

Jacob, S., Hayreh, D. J. S. and M. K. McClintock. 2001. 'Context-dependent effects of steroid chemosignals on human physiology and mood'. **Physiology and Behavior**, vol. 74. 15-27.

Jarymowicz, M. 2007. 'Czy emocje mogą się zmienić w polu świadomości'. In Ohme, R.K. (ed.). **Nieuświadomiony afekt**. Gdańsk: GWP. 17-28.

Ji, S. 1997. 'Isomorphism between cell and human languages: molecular, biological, bioinformatic and linguistic implications'. **BioSystems**, vol. 44. 17-39.

Jibu, M. and K. Jasue. 2004. 'Quantum brain dynamics and quantum field theory'. In Globus, G. G., Pribram, K. H. and G. Vitiello (eds.). **Brain and being. At the boundary between science, philosophy, language and arts**. Amsterdam/Philadelphia: John Benjamins. 267-290.

Kiklewicz, A. and J. Dębowski (eds.). 2008. **Język poza granicami języka**. Olsztyn: Instytut Dziennikarstwa i Komunikacji Społecznej UWM.

Kiklewicz, A. 2006. **Język, komunikacja, wiedza**. Mińsk: WTAA Prawo i Ekonomika.

Kitto, K, Ramm, B., Sitbon, L. and P. Bruza. 2011. 'Quantum Theory beyond the physical: information in context'. **Axiomathes**. 21. Springer: published online 14 January 2011. 331 -345.

Krings, H. 1986. 'Translation problems and translation strategies of advanced German learners of French'. In House, J. and S. Blum-Kulka (eds.). **Interlingual and intercultural communication. Discourse and cognition on translation and second language acquisition studies**. Tübingen: Günter Narr Verlag.

Kuhn, T. 1970. **The structure of scientific revolutions**. Chicago: University of Chicago Press.

Kussmaul, P. 1995. **Training the translator**. Amsterdam: John Benjamins Publishing Company.

Lakatos, I. 1978. **The Methodology of Scientific Research Programmes: Philosophical Papers**. Vol. 1. Cambridge: Cambridge University Press.

Lamb, S. M. 1999. **Pathways of the brain**. Amsterdam: John Benjamins Publishing Company.

Langacker, R. W. 1987. **Foundations of Cognitive Grammar**. Vol. 1. Stanford: Stanford University Press.

Lörscher, W. 1991. **Translation performance, translation process, and translation strategies. A psycholinguistic investigation.** Tbüingen: Günter Narr Verlag.

Mandler, G. 1984. **Mind and body.** New York: WW Norton and Company.

Maruszewski T. and E. Ścigała. 1998. **Emocje, aleksytymia, poznanie.** Poznań: Wydawnictwo Fundacji Humaniora.

Miller, S. 2008. **Awesomism! A new way to understand the diagnosis of autism.** Bloomington: iUniverse.

Mühlhäusler. P. 1995. 'Metaphors others live by'. **Language and communication.** Vol. 15. No 3. 280-288.

Mühlhäusler. P. 2003. **Language of environment. Environment of language.** London: Battlebridge.

Ohme, R.K. 2003. **Nieuświadomiony afekt.** Gdańsk: GWP.

Ohme, R.K. 2007. **Podprogowe informacje mimiczne.** Warszawa: Wydawnictwo PAN; SWPS.

Patton, M. Q. 2002. **Qualitative research and evaluation methods.** Thousand Oaks: Sage Publications.

Pause, B. M. 2004. 'Are androgen steroids acting as pheromones in humans?'. **Physiology and Behavior**, vol. 83. 21-29.

Penrose, R. 1995. **The shadows of the mind.** London: Vintage.

Penrose, R. 2004. **Droga do rzeczywistości.** Warszawa: Prószyński i S-ka.

Plotnitsky, A. 2004. The unthinkable: nonclassical theory, the unconscious mind and the quantum brain. In Globus, G.G., Pribram, K.H. and G. Vitiello (eds.). **Brain and being. At the boundary between science, philosophy, language and arts.** Amsterdam/Philadelphia: John Benjamins. 29-45.

Prete, N. 2004. 'Doubling the image to face the obscenity of photography'. In Globus, G. G., Pribram, K. H. and G. Vitiello (eds.). **Brain and being. At the boundary between science, philosophy, language and arts.** Amsterdam/Philadelphia: John Benjamins. 1-19.

Pribram. K. H. 1. 'An instantiation of Eccles brain/mind dualism and beyond'. http://www.paricenter.com/center/ DOA: 18.10.2011r.

Pribram, K. H. 2. 'Consciousness reassessed'. http://www.paricenter.com/center/ DOA: 18.10.2011r.

Pribram, K. H. 3. 'Brain and mathematics'. http://www.paricenter.com/ center/ DOA: 18.10.2011r.
Pribram, K. H. 2004. Brain and mathematics. In Globus, G. G., Pribram, K. H. and G. Vitiello (eds.). **Brain and being. At the boundary between science, philosophy, language and arts.** Amsterdam/Philadelphia: John Benjamins. 215-239.
Puppel, S. (ed). 1995. **The biology of language.** Amsterdam: John Benjamins.
Puppel, S. 1996. **A concise guide to psycholinguistics.** Poznań: Bene Nati.
Puppel, S. 1999. 'Psycholinguistics and the foreign language teacher'. **Acta Neophilologica I.** Olsztyn: Wydawnictwo UWM. 73-83.
Puppel, S. 2006. 'Czarna Skrzynka Noama Chomsky'ego a trychotomiczny model umysłu: perspektywa psycholingwistyczna'. Poznań: Adam Mickiewicz University, Department of Ecocommunication. Electronic journal **Oikeios Logos** 2.
Puppel, S. 2008. 'Communicology: remarks on the reemergence of a paradigm in communication studies'. In Puppel, S. and M. Bogusławska-Tafelska (eds.). **New Pathways in Linguistics.** 11-22.
Puppel, S. 2011. 'Human communication and communicative skills: a general philosophy and evolving practical guidelines'. In Puppel, S. and M. Bogusławska-Tafelska (eds.). **New Pathways in Linguistics.** 107-118.
Radden, G. and R. Dirven. 2007. **Cognitive English Grammar.** Amsterdam/Philadelphia: John Benjamins.
Rantala, M.J. Eriksson, C.J.P., Vainikka, A. and R. Kortet. 2006. 'Male steroid hormones and female preference for male body odor'. **Evolution and Human Behaviour**, vol. 27. 259-269.
Reiss, S. 2001. 'Secrets of happiness'. **Psychology Today.** Vol. 1. 50-56.
Rossano, M. J. 2011. 'Setting our own terms: how we use ritual to become humans'. In Walach, H., Schmidt, S. and W. B. Jonas (eds.). **Neuroscience, consciousness and spirituality.** London/New York: Springer. 39-55.
Sale, J., Lohfeld, L. and Brasil, K. Revisiting the quantitative-qualitative debate: implications for mixed –methods research. **Quality and Quantity.** Vol. 36. 2002. 43-53.

Samuelsen, C.L. and M. Meredith. 2009. 'The vomeronasal organ is required for the male mouse medial amygdale response to chemical-communication signals, as assessed by immediate early gene expression'. **Neuroscience**. Vol. 164. 1468-1476.

Schaefer, V. 2006. 'Science, stewardship, and spirituality: the human body as a model for ecological restoration'. **Restoration Ecology**, vol. 14, no 1. 1-3.

Schrödinger, E. 2007. **What is life? With mind and matter and autobiographical sketches**. Cambridge: CUP.

Searle, J. R. 2000. 'Minds, brains and programs'. In Cummins, R. and D.D. Cummins (eds.). **Minds, brains and computers. The foundations of cognitive science**. Oxford: Blackwell. 140-152.

Sills, L. and J. Lown. 2008. 'The field of subliminal mind and the nature of being'. **European Journal of Psychotherapy and Counselling**. Vol. 10. No 1. 71-80.

Stamenov, M.I. 2004. 'The rheomode of language of David Bohm as a way to re-construct the access to physical reality'. In Globus, G. G., Pribram, K. H. and G. Vitiello (eds.). **Brain and being. At the boundary between science, philosophy, language and arts**. Amsterdam/Philadelphia: John Benjamins. 148-164.

Stockhorst, U. and R. Pietrowsky. 2004. 'Olfactory perception, communication, and the nose-to-brain pathway'. **Physiology and Behaviour**, vol. 83. 3-11.

Summhammer, J. 2011. Quantum cooperation. **Axiomathes**. Springer: published online 31 December 2010.

Tirkkonen-Condit, T. S. (ed.). 1991. **Empirical research in translation and intercultural studies**. Tübingen: Günter Narr Verlag.

Tudor, I. 2001. **The dynamics of the language classroom**. Cambridge: CUP.

Walach, H. and N. von Stillfried. 2011. 'Generalised quantum theory – basic idea and general intuition: a background story and overview'. In **Axiomathes**. Vol. 21. 185-209.

Walach, H., Schmidt, S. and W.B. Jonas (eds.). 2011. **Neuroscience, consciousness and spirituality**. London/New York: Springer.

Walach, H. 2011. 'Neuroscience, consciousness, spirituality – questions, problems and potential solutions: an introductory essay'. In Walach, H., Schmidt, S. and W.B. Jonas (eds.). **Neuroscience, consciousness and spirituality**. London/New York: Springer.

Wilson, R. A. and F. C. Keil (eds.). 1999. **The MIT encyclopedia of the cognitive sciences**. Cambridge, Mass.: The MIT Press.

## Internet sites

http://juicylivingtour.com/2011/10/25/the-unified-field-nassim-haramein/ DOA: 26.10.2011.

http://southerndenmark.academia.edu/SuneSteffensen/Papers/1055129/Language_ecology_and_society_an_introduction_to_Dialectical_Lingu

http://theresonanceproject.org/ DOA: 26.10.2011.

http://www.brucelipton.com/ DOA: February 2012.

http://www.youtube.com/watch?v=eMlAdEr45dY

http://www.youtube.com/watch?v=vHpTYs6GJhQ DOA: 17.10.2011.

Interview with dr Amit Goswami http://www.youtube.com/user/liloumace?blend=1&ob=videomustangbase#p/u/49/DDaYKIhky3Q

Interview with dr Fred Allan Wolf: http://www.liloumace.com/Dr-Quantum-Fred-Alan-Wolf-PhD-Time-Space-Matter-Quantum_a1790.html

http://www.youtube.com/watch?NR=1&feature=endscreen&v=0Eeuqh9QfNI Stanford University lectures on Quantum entanglement

http://www.youtube.com/watch?v=ERlOJAaLxG0 Uniwersytet Śląski: lectures on quantum mechanics

**Interfaces**
**Bydgoszcz Studies in Language, Mind and Translation**

Edited by Anna Bączkowska

Volume 1   Tomasz Ciszewski: The Anatomy of the English Metrical Foot. Acoustics, Perception and Structure. 2013.

Volume 2   Marta Bogusławska-Tafelska: Towards an Ecology of Language, Communication and the Mind. 2013.

www.peterlang.de